FOOD PRODUCTION AND INDUSTRY

Edited by **Ayman Hafiz Amer Eissa**

Food Production and Industry
http://dx.doi.org/10.5772/59527
Edited by Ayman Hafiz Amer Eissa

Contributors
Gisela Maria Dellamora Ortiz, Ângelo Samir Melim Miguel, Tathiana Souza Martins Meyer, Daniel Ernesto Rodriguez Fernandez, Makoto Kanauchi, Sakiko Hatanaka, Makoto Shimoyamada, Mehdi Zahaf, Elisa Hirooka, Fabio Luis Silva Faria-Oliveira, Rogélio L. Brandão, Ieso M. Castro, Margarete Saraiva, Fernanda Piló, Raphael Hermano Santos Diniz, Hygor Mezadri, Fernanda Godoy-Santos, Suzana Caetano Da Silva Lannes

Published by InTech
Janeza Trdine 9, 51000 Rijeka, Croatia

Publishing Process Manager Iva Simcic
Technical Editor InTech DTP team
Cover InTech Design team

First published October, 2015
Printed in Croatia

Additional hard copies can be obtained from orders@intechopen.com

Food Production and Industry , Edited by Ayman Hafiz Amer Eissa
p. cm.
ISBN 978-953-51-2191-6

For EU product safety concerns:
IN TECH d.o.o., Prolaz Marije Krucifikse Kozulić 3, 51000 Rijeka, Croatia,
info@intechopen.com or visit our website at intechopen.com.

Meet the editor

Prof. Ayman Amer Eissa received his B.Sc. in 1986 and M.Sc. in 1992 from Minoufiya University, Egypt, and his Ph.D. from Martin Luther University, Germany, and Minoufiya University (channel system) in 1999. He completed fellowships in biosystems engineering and quality control in 2007 from Germany. Prof. Amer Eissa previously served as the director of the Center of Marketing Service University, Minoufiya University. He was previously professor of food process engineering in the Department of Agricultural Systems Engineering, College of Agriculture and Food Sciences, King Faisal University, Saudi Arabia. He is currently professor of food process engineering at Minoufiya University, Egypt. His research is directed at machine vision processing and development of different package systems and transportation for food products. Prof. Ayman Amer Eissa is the author or co-author of over 50 journal articles and the author, co-author or editor of more than five books. He held memberships in different honorary societies of food engineering. He has been a technical reviewer for many journals on food engineering. He has supervised over 30 M.Sc. and Ph.D. students.

Contents

Preface

This book is an example of a successful and excellent addition to the literature on the topic of Food Production and Industry within the scientific world. The book is divided into six chapters, consisting of selected topics in food production and consumption and food preservation. All the six chapters have been written by renowned professionals working in Food Production and Industry and related disciplines.

The first chapter of the book deals with local food research, an area of study with a vast number of possible areas of future research. Local farmers will find value in knowing that market potential does exist for their product, and consumers are expressing an interest in purchasing locally produced food at all channels of distribution. Their motivation to buy local food products is not driven by fear and concern over food products but rather by key information on country of origin, nutritional value, freshness and healthiness. This information can be found in the packaging (i.e. date of packaging, contact information, etc.). In terms of channels of distribution, many people expressed a willingness to purchase local food products if available at their supermarkets, local grocery stores and specialty stores, while only a very small percentage considered the farmer's market. However, this may defeat the philosophy of buying and eating local. Further, it is important for farmers to note that consumers do express a willingness to pay a premium for their local food products.

The shelf life of foods is being reformulating based on scientific evidence to the satisfaction of health authorities, import countries and different stakeholders. For this, a systematic approach for establishing by-product category shelf lives should be adopted. Research and technological centres, universities, food industries and sectorial associations should agree in standardized protocols to face this challenge. This could be applied to food production systems where a complex set of variables requires monitoring combined with complex reasoning to assure safety and quality.

The remaining chapters in Part I deal with commercial applications of food production and consumption. From our point of view, as consumer safety questions are discussed, food preservation through radiation as a contribution to food safety will reach the same recognition as the sterilization of medical products in terms of preventing the spread of infectious diseases. For this reason, scientists have the responsibility to help the consumer understand the radiation process and its potential to protect our health and thus improve our lives.

Part II consists of three comprehensive overviews: The Role of Yeast and Lactic Acid Bacteria in the Production of Fermented Beverages in South America; Safety and Quality in Agricultural Product Chain in Brazil; and Low Fat Foods.

We truly hope that this book, with its visionary approach, will be a valuable addition to the food engineering literature and will promote interest in Food Production and Industry research, development and implementation. Finally, I acknowledge that each of the authors has provided their extraordinary competence and leadership in the specific field and that the publisher, with its enterprise and expertise, has enabled this project which includes various nations and continents. Thanks to them, I have the honour to be the editor of this book.

Prof. Dr. Ayman Hafiz Amer Eissa
Professor of Food Process Engineering,
Department of Agriculture Engineering,
Minoufiya University,
Shibin El-kom, Egypt

Section 1

Food Production and Consumption

Chapter 1

Alternative Foods — Marketing Perspectives on the Production and Distribution Systems

Mehdi Zahaf

Additional information is available at the end of the chapter

http://dx.doi.org/10.5772/60565

Abstract

Most societal and environmental challenges that humanity is currently facing relate to unsustainable consumption patterns and lifestyles. This unsustainability is supported by consumerism and producerism that have high interests for fierce productivity. This study presents a supply-demand marketing perspective based on the current literature and current market realities. Local food production and consumption are topics gaining much attention in academic marketing research and marketing practice. Local food systems regenerate the relationships between producers and consumers. A mixed design is used to assess the added values of the production and distribution systems in Canada. On one hand the supply side situation was assessed by conducting personal in-depth interviews with alternative food producers and market intermediaries. On the other hand, a survey was administered to local food consumers to assess their consumption patterns. Findings show that producers/farmers have a number of positive outcomes. Intrinsic values such as taste, quality and freshness, nutrition and health, are very important to the consumer both for local food products in general and local organic foods products in particular. Further, it is determined that price is important but not determinant of local foods purchase. Further, consumers' issues and concerns over conventional food products are determinant factors when purchasing local food products. It is also important to realize that consumers may not fully understand the meaning of locally produced food.

Keywords: locally produced foods, purchasing patters, food concerns, value delivery, food chain

1. Introduction

Local food production and consumption are topics gaining much attention in both current academic marketing research and marketing practice. Global food systems are being criticized due to environmental, social and economic factors, which can be explained using the concept of local farming and local food culture. This is often referred to as embeddedness. Embeddeness is a core concept in food system research; it has been borrowed from economic sociology, and is widely used to characterize the different types of food systems. At one end, the dis-embedded globalized system is characterized by "industrial food" and not well-informed food choices. At the other end, the embedded localized system is described as a "human" and social system where food is produced and eaten locally [23]. As consumers express fears and concerns about the global food system, they are beginning to seek food alternatives. This in turn results in the emergence of new production options such as Farmer's Markets, Community Supported Agriculture groups, and community / home gardens. Country of origin used to be key in consumers' purchasing decision, but now food mileage has taken over it. People are beginning to care about the – local - origin of food products and are seeking locally produced foods when making their purchase decisions. For instance, the growing popularity of local food diets, such as the 100 Mile Diet is an indicator of a new trend towards consumption of local products.

As the population gains awareness, expresses concerns about global the system and demands the availability of locally grown food products, an opportunity may exist for farmers to target this market and modify production as necessary. Farmers would benefit from understanding consumer views on locally produced foods, and if market potential exists, then be able to formulate a marketing approach in line with consumer demands. Currently, there is limited research on the topic of consumer perception of alternative foods, and even less on locally produced food. As a matter of fact, from an academic research perspective it is important to understand the evolution of food research. Most of the research cited in the current study is fairly current, indicating that this local food research is an emerging area of study with remarkable potential. While interest in "buy local" food campaigns has grown in recent years, academic research in this area is divided along two distinct disciplinary lines. Rural sociology and rural geography examine food from a production standpoint, whereas marketing research and consumer behavior focus more on consumption [25]. In addition, there is need to consider a new body of research on local food products in relation to the concepts of embeddedness, sustainability, and their interconnectedness.

This chapter presents a supply-demand marketing model based on the current literature and current market realities. The model integrates several key factors and is built to answer practical questions related to the consumers' lifestyle, awareness of food alternatives, and motives to be active societal consumers.

2. Food system economics

2.1. Local versus global food systems

The food system is composed of all activities involving production, processing, transport and consumption of food [10]. Food systems are categorized in local food systems and global food systems.

Increased globalization of food systems, large-scale production and distribution, and retail sales have changed the way food is produced and consumed. Prior to World War II, most of the agricultural production and marketing systems had strong local or regional bases. Since that time, many farms evolved to embrace new technologies to gain competitive advantages in the marketplace. As farming practices evolved and farmers became more specialized, the number of farms in both the U.S. and Canada fell significantly. [19] describes how American consumers have continued to demand quality food, but at the lowest price possible. The result of such demand has been the relocation of large farms to where land and labor costs are low. This system encourages a divide between land and people, between farm and city, and producers/farmers and consumers; food is no longer a social activity.

Nowadays, the idea of committing to eating only locally produced foods has become popular. Further, local agricultural knowledge and traditional farming methods have been lost and replaced with standardized production methods and consumption patterns. In this complex globalized food system, most food production and processing occur sizeable distances away from where people live, and buy groceries. *"Today's global food system contains fewer farmers on larger farms producing foods for distant markets."* [19]. In their research, [7], reference the work of McMichael (2000), Kimbrell (2002), Nestle (2006), Pollan (2006, 2008), indicate that the increasingly industrialized and globally extensive food systems of today are constantly being criticized regarding health, social, economic and environmental problems. This disconnect has also created many concerns with respect to food safety, food security, health, and sustainability. Local alternatives that are developing, as some would argue, are a response to this globalization trend [16]. Hence, strong interest by consumers and farmers to reconnect and re-establish these connections has led to an increase in the number of farmers markets, community supported agriculture groups, and local food initiatives such as food baskets.

2.2. Food security perspective

"Food security exists when all people, at all times, have physical and economic access to sufficient, safe and nutritious food that meets their dietary needs and food preferences for an active and healthy life" [27]. Most societal and environmental challenges that humanity is currently facing relate to unsustainable consumption patterns and lifestyles. Consumption as it relates to this vision does not simply fulfill needs, but also satisfies new wants and desires. This unsustainability is supported by consumerism and producerism that have high interests for fierce productivity. Food is no exception to this global trend. The severe industrialization of the agricultural system is making the new food chain rely on pesticides and antibiotics. Processed and transformed, chemically-preserved, and nutritionally-poor, food products are nowadays more accessible

and convenient options to healthy food. As a matter of fact, the results of this unsustainable production system are disastrous. It has contributed to the degradation of the conventional agriculture, and overall public health. In fact, the last two decades were driven by two major trends in the agriculture industry; first, an increase in the use of genetically modified food (GMO); and an increase in food-related diseases, such as mad cow, bird flu, and more recently the horsegate.

Emerging efforts to provide food safety and quality has led to growing number of national and international quality assurance schemes. To this end, several "new" alternatives eliminate a number of concerns towards industrial food production and distribution. These "alternative food" options include labels such as local, natural, pesticide-free, ecologically friendly, slow food movement, and localvores. A *localvore*[1] is a person who eats foods grown locally; within often defined as within a 100-mile radius. In eating locally, most localvores hope to better understand their food sources and support their local economy. Advocates of these movements are against any industrialization of the food chain, its production and distribution. Three pillars are key here: food mileage, support for the local economy and production methods. It is clear that these new "sustainable" alternatives are based on food security principles: food access, utilization and stability [6].

2.3. Embeddedness and sustainability

Embeddeness is a concept widely used in food system research. Borrowed from economic sociology, the concept of embeddedness is widely used to characterize the different types of food systems: the dis-embedded globalized system and the more embedded localized system [23]. The embeddedness concept can be used to enrich the understanding of the broader sets of values tied to consumer motivations to buy at farmers markets [7]. As a matter of fact, local markets, where people exchange goods and services, are embedded in a network of social relations and meaningful systems of norms and rules [16]. Central to local food systems is the face-to-face connections and interactions between producers and consumers. Global markets are shaped less by social networks and norms, and impose instead compliance with their own norms of efficiency, rationality, optimizing and 'time saving' behavior. This global food system is experiencing societal, cultural, and moral disembedding. Ethical embeddedness is also being criticized resulting in a growing sense of distrust. Perhaps the development of more local systems is a result of social theory. This theory suggests that in situations of increasing distrust, alternative movements will emerge as consumers get organized to overcome their sense of insecurity [16]. As a result, alternative food schemes such as farmers' markets and community supported agriculture have emerged giving consumers access to food they perceive to be less harmful to the environment, society, and health.

The long term stability and sustainability of food systems is an interesting area of research. Academic literature on food research indicates a renewed interest in sustainable agriculture and food systems [4]. Sustainability has been described as having three conceptual pillars: social, environmental and economic factors. A sustainable food and agriculture system

---

1*local = place, vore = to consume*

http://dx.doi.org/10.5772/60565

requires economically viable farms and productive farmland to provide quality food [4]. Locally grown foods and farmers' markets contribute to the sustainability of the food system in many ways. Significant job and income impacts resulting from increased consumption of locally grown products are natural by-products of such systems. Ancillary benefits including increased access to healthy foods, business opportunities and drawing customers to downtown areas are also new trends.

3. The marketing of alternative foods

3.1. Food production, operations size and distribution logistics

It is worth noting that the alternative food *movement*, which began as an alternative style of production among small farms looking to reduce their environmental footprint and to differentiate their products from commercially produced foods, is going mainstream. Production methods, operations size, and distribution are key here. To this end sustainable food production *should* combine best environmental practices, preservation of natural resources, animal welfare standards while ensuring no use of genetic engineering, pesticides, additives, or fertilizers. Overall, sustainable alternative production systems deal with small and controlled size farming. This limits the distribution of such productions to the local or regional markets. At this time production of such farms is rather limited amounting to a few hundred tons. Such a volume will be of little interest to mainstream grocery chains. Distribution logistics and costs would consume a significant part of the producers/farmers margins.

Industrial farming addresses efficiently and effectively the challenges related to the cost and logistics of moving produced foods to national and global markets. Conventional food value chain applies an important downward pressure on price leading to the issues of profitability and productivity. As a consequence, some alternative foods such as organics moved slowly toward large-scale operations. This has resulted, for some small farmers concerned with the philosophical aspects of organic production, in diminished credibility of the organic standard and a refusal to participate. It has also hardened the value chain against entry by these small farmers. Hence, the challenge that the alternative food system is facing is a gap that spans between the consumerism/producerism system in place, the current food chain, and the alternative value delivery network/value chain. Therefore, distribution flows and logistics should be explored and studied.

In conventional food systems there exists between producers and consumers of food products a series of handlers involved in the processing and distribution at various stages of the journey from farm to plate. Traditional retail, with its focus on profit, seeks consistent supply of products. Imports from warmer climates offer this consistency. We see California and Mexico lettuce occupying shelves year-round because, for reasons of efficiency, retailers prefer to deal with a single supplier rather than displace the year-round supplier with a seasonally-available product. This is another major factor, and one that could mitigate the seasonality of foods. Large-scale farming is sustained by important economies of scale while small-scale farming involves higher production costs due to limited operations leading to higher selling prices. As a result, there is a wide variety of product classifications depending on the production methods and thus, the operations size. This in turn gives raise to 2 distinct distribution systems. Long channels, eg. retail chains, that add value through price and high distribution intensity, and short channels, eg. direct from producers, that add value through their production methods and sustainable practices. Conventional distribution channels, characterized by a longer channel where consumers do not see and interact with the producer and where the information about food is limited, is targeted toward consumers that look for a one-stop grocery shopping experience [11, 12]. Unlike organic foods, that are observing a decline in consumers' trust, alternative food options are using mechanisms based on small-scale operations and direct channel distribution. Direct channels such as the farmers market is targeted toward consumers that look to interact – socially - with the producers [22], ask them questions about their production methods, food origin and variety, and cooking tips. Distinct trends are thus observed in the alternative food distribution.

3.2. Consumer marketing theory and alternative foods

In an environment of distrust toward the new food chain, addressing the issue of unsustainability from the demand side by ensuring product quality and product knowledge, labeling or setting proper pricing and communication strategies does not seem to be enough [11, 12]. Conversely, building trust in the food supply requires tools such as quality and non-profitability-based production processes.

The traditional marketing approach and specifically consumer marketing theory are not sufficiently prepared to handle the advent of new types of consumers such as organic food (OF) consumers, localvores, or slow food movement consumers. These consumers are looking for more than a product, i.e., they are looking for value products [1]. Moreover, the literature to date is of great interest but offers little insight into the motives to buy alternative food products. Growing consumer demand for alternative food products has been attributed to consumers' concerns regarding nutrition, health, the environment, and the quality of their food [8, 14]. Further, various studies conducted in Europe and the US have explored the consumer behavior and have tackled the issue of determining consumers' motivations and preferences for organic food products and local food products [26, 28]. Although some consumers are environmentally conscious, most studies confirm the predominance of egocentric values like health, attitude towards taste, and freshness that influence alternative food choices [8, 28].

Conversely, studies dealing with the concept of environmentally friendly products represent a mature research field covering different aspects of the marketing process namely, consumer behaviour, marketing mix, and marketing strategies. It is also interesting to shed the light on similarities and differences between green products and alternative food products. The issue of psychographic and personal variables in consumer behaviour dealing with environmental or green products has been dealt with by many researchers [2, 3, 5]. Hence, some of the variables that might shape the buying behavior of alternative food products are (i) perceived consumer effectiveness: consumers' believe that the efforts of an individual can make a difference in the solution to environmental problems [2, 3, 5], (ii) ecological concern: consumers' level of ecological concern is related to their willingness to purchase green products, (iii) locus of control: refers to consumers' perception about the underlying main causes of events in their behavior [21], and (iv) faith in others [3].

3.3. Motivations to buy local

Consumers make personal decisions on food choices every day. Consumer preferences can have both extrinsic and intrinsic values. For example, extrinsic values would include price and brand, whereas intrinsic would include taste, quality, and freshness. In reviewing literature on why individuals buy local food products, there appears to be both private needs and altruistic intentions that motivate an individual to buy local. The three main benefits of local food systems are the environmental, economic and social benefits received. *Environmentally,* local food products' sales reduce greenhouse gases because the food has reduced transportation miles (reduced carbon footprint). This also results in better food quality and fresher food availablility. *Economically,* local food systems help create local jobs, stable regional employment and provide better return on products (40-80% increase) through direct marketing at farmers' markets. *Socially,* key benefits are achieved through greater trust and connectedness between consumers and farmers. Many consumers are interested in local food products because of the perceived benefits of freshness, taste, and quality [18]. Because local products can potentially reach their consumers quicker, there is also a perception of a higher nutritional content.

In a study of food marketing consultants (Tellstrom et al. (2006), local food is described as "*an invention to reflect what urban consumers perceive about the idyllic rural countryside*". A common theme however, is the importance of social interactions, as well as food freshness and the ability to support local farmers. Several other factors come into play such as food safety, quality, and health benefits. There is also evidence that some consumers are making civic choices and society-focused statements with their purchase decisions [17]. Rural sociologists argue that some local food consumers are not trading off and transacting on products, but rather they are seeking shared product values and engaging in reciprocal and trustful relationships with farmers and food producers [25]. According to an IPSOS Reid Survey of 2006 at Ontario Farmers Markets revealed that 95% of shoppers felt that "*buying products produced in your community is either very important or moderately important*". More importantly, 91% of the Ontarians surveyed indicated they would buy locally produced food if they could find it in their grocery stores.

4. Research design

4.1. Objectives

The growing popularity of local food marketing creates an interesting area of new research. Whereas the majority of previous research is focusing on the demand side, this research aims to bridge the demand to the supply by uncovering bilateral factors affecting the marketing of alternative foods. Further, it is crucial to understand the motives and purchasing patterns of alternative food consumers. This helps to understand the marketing actions and logistics underlying the production and distribution systems of local food products.

Hence, the objectives of the research are:

- Assess the importance of local foods production and distribution systems in creating and shaping the local foods market;
- Understand the motives and concerns, product preferences, and consumption patterns of local food consumers; and
- Build a local food products marketing model.

4.2. Conceptual framework

To address the abovementioned objectives, a conceptual framework has been developed (cf. Figure 1).

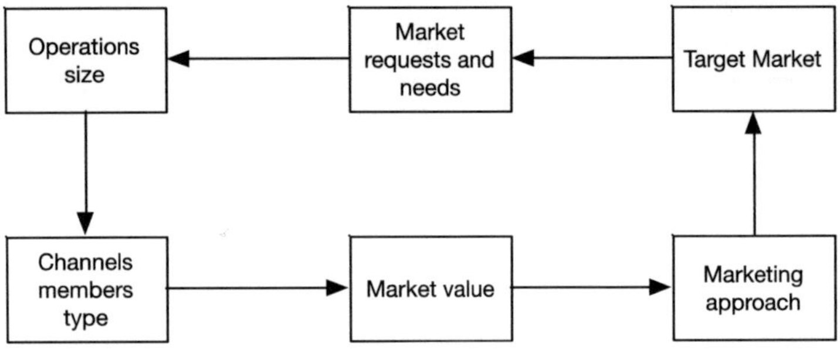

Figure 1. Conceptual Framework

It shows the value delivery network that will be assessed in this study. Basically consumers are assumed to have requests regarding food quality, freshness, environmental and economic impacts, and healthiness. This is assumed to depict a certain size of operations (large versus small). This in turn will impact the type of channels member involved in these operations. It is assumed here that the distribution channels are very short, counting a maximum of 2 members: 1 producer/farmer and 1 distributor (if there are any). These channels create certain values that are logically different depending on the point of sale. Lastly, depending on the market coverage and the channel size, farmers, producers, or distributors will have a marketing approach adapted to the value offered to the target market.

4.3. Design and procedure

To address the abovementioned objectives, a mixed design has been used to assess the added values of the production and distribution systems in Canada. The supply side situation was assessed by conducting personal in-depth interviews with alternative food producers and market intermediaries. 60 in-depth interviews were conducted in Ontario and Quebec, Canada, in Fall 2013 and Winter 2014. These interviews were based on an interview guide that lasted about 30 to 45 minutes. The guide probes respondents in the local food products value delivery network to discuss the actual structure of their distribution channel and their marketing approach. The interview guide is composed of three sections. The first section deals with the marketing approach used by producers/distributors, the second section aims to explore the structure of the channel of distribution, and the last section deals with how producers/distributors perceive the market structure. This three-prong data collection scheme helps to determine with minimal bias the characteristics of this new market from a supply side perspective. Distributors were approached according to (i) market size, (ii) local food categories produced/sold, and (iii) by channel position (producers versus distributor). The in-depth interviews were recorded, transcribed, coded, and analyzed using content analysis (cf. [13]. This technique allows to process large amounts of textual information, systematically identifies its properties, and determines the frequencies of most used keywords.

On the other hand, to explore the local food products consumption patterns, a large-scale survey was administered to 1752 consumers. The population targeted for this study is alternative food shoppers. These consumers had to buy local food products and were randomly surveyed at small producers' farms gate, community farmers, farmers' market, community groceries, specialty stores, and community chain stores. The following food categories were targeted: fruits, vegetables, dairy, and meat. To this end and to test the 2nd objective, a structured questionnaire was designed. This objective here is twofold: (i) to determine the impact of consumer concerns and motivations on local food product purchase, and (ii) to study product attributes preferences and more generally the consumption patterns of local food consumers. The survey was developed by selecting other case study questionnaires on the topic of locally grown products [9, 20]. Section 1 of the survey deals with consumers concerns over local food products in general, and assesses their consumption, preferences and purchasing patterns. These questions determine purchasing frequency, typical shopping habits, spending, product attributes, and concerns over local food products. Section 2 of the survey deals with consumers perceptions and willingness to buy locally produced food. The last section of the survey is used to determine the socio-demographic profiles of the respondents including variables such as gender, occupation, age, number of people in the household, education, income and postal code. Prior to administering the survey, a pre-test was done and minor modifications were made. Quantitative data for this study was analyzed using SPSS statistical package. Data was cleaned and missing values were replaced using the mean. Out of the 1752 collected surveys, 1611 questionnaires were usable. A combination of reliability tests and factor analyses were performed to determine the reliability and the structure of the data.

5. Results

5.1. Production and distribution of local foods

5.1.1. Themes generation

In-depth interviews took place in Ontario and Quebec, Canada. It targeted four cities selected as follow: 1 city and 1 community in each province, and 15 channel members in each location for a total of 60 in-depth interviews. The following channel member categories were targeted: small producers, community farmers, farmers' market, community groceries, specialty stores, and community chain stores. This provides a representative image of the different distribution systems based on city size and culture (French and English) and helps to uncover how producers/distributors develop marketing approaches to target this market niche. Content analysis of in-depth interviews was conducted. Results of the interviews analysis generated four main theme categories: local foods market, local foods definition, marketing issues, and strategies/approaches developed by distributors/producers (cf. Table 1). All themes are derived from the first research objective and clearly reflect the spirit of the interviews.

Theme categories	Theme	Definition
Local foods market	Trend	New type of consumers: localvores
	Market forces	Supply Driven by demand
	Education	Consumer's education about benefit of alternative foods
Local food definition	Local	Local foods product most important attributes
	Definition	Confusion regarding organic/natural/local
Marketing issues	Price	Pricing issues related to local food production
	Distribution	Main store categories where local food products are usually sold and channel length
	Product	Local foods product most important attributes
Strategies and approaches	Communication	One to one relationship or individual marketing
	Network	System of action to develop and enhance the offering

Table 1. List of Generated Themes

5.1.2. Local foods value delivery

Farmers, producers, and distributors have consistent perceptions of the distribution system and the distribution structure. The respondents interviewed for this study generally agree that the local foods market is growing slowly with a solid base of knowledgeable consumers. This market shows also substantial opportunities. More specifically they mentioned an increasing diversification of local food products and a conservative expansion in the type of distribution channels. This new type of offering is mainly derived from the "failure" of the conventional foods system to offer embeddedness. Producers and distributors argue – almost unanimously – that the market is asking for alternative way of producing, distributing and eating. This makes more sense when considering that the organic market went mainstream and failed on several aspect of the value chain and value delivery network.

Arguments put forward by these respondents were related mainly to (i) food mileage and (ii) industrialization of the distribution. Organic food products have failed to carry out value mainly on the second dimension, namely industrialization, and sometimes on both dimensions

http://dx.doi.org/10.5772/60565

(food mileage and industrialization). Further, all interviewees from various channels agree to say that consumers are becoming more educated and make smarter food choices. However, it is clear that there are fundamental differences in their purchasing and consumption patterns. Short channels members, such as farmers selling at farms gate, stated that consumers buying at their point of sale are very knowledgeable and are looking for more than a standard purchase. The derived value is based on building a long lasting relationship with the farmer/ producer. Conversely consumers buying from more conventional channels, such as specialized stores, are looking for a different purchasing and consumption value. They are also looking for convenience.

Depending on the type of distribution channel considered, findings clearly highlight factors and dimensions that influence consumers' trust when making local food products purchases.

Specialty Stores: As far as specialty stores perspective goes (independent stores as well as community specialty chain stores), the corresponding market shows limited growth in terms of variety, price and quality. These stores are able to provide consumers with a relatively larger variety and convenience whereas larger grocery stores are not differentiating their offering. This is mainly related to the value offered in these longer channels: price is crucial; quality being downgraded for competitive economies of scales. Specialty stores managers indicate that it is obvious and clear that the product label is important. Target consumers feel very comfortable knowing what to buy and finding all information they look for. Conversely, the brand name is not important. Consequently, consumers are not looking for an *industrial* or *conventional* value, rather they are looking for a – convenient – *embedded* value.

Community Grocery Stores: For community grocery store managers, consumers are less inclined to buy local food branded products or packaged local food products. This is probably more related to the combination of embeddedness, sustainability, and food mileage. Conversely, some consumers are explicitly looking for local food products, as they know that store managers are more approachable. This is a guarantee for quality and counterbalances the lack of brand effect. Here, the value is mainly based on the relationship with the manager.

Producers/Farmers Market: When it comes to producers and farmers' markets, interestingly enough, all of them acknowledge that consumer do not trust any labels. They say that consumers would like to get information regarding the products they buy but when they interact with the farmers/producers and discuss the production methods with them, they build a trust relationship that offsets the label value. Hence, the value is based on the production methods. Lastly, most producers and farmers are increasing their local food products sales using traditional sales approaches to market their offering. Farmers and producers do not like to talk about marketing of local foods; this has an *industrial* connotation. Further, being able to expand supply is a big dilemma that translates into controlled supply and limited growth rates. With the growth in popularity of the slow food movement and organic food products, community producers and farmers with small to medium size farms, are managing and controlling this new market niche. It is becoming a sustainable and fair source of revenue. Further, several farmers and producers are selling at farms gate, which considerably lowers their selling price; distribution cost being reduced to zero.

To recapitulate, there is a thin distinction between local food products producers and distributors in terms of production and distribution processes, marketing approaches, and market perception of the value offered. Brands definitely do not add any value whatever the type of distribution channel whereas the product label may play a substantial role for some distributors. Furthermore, store managers as well as the production methods contribute in creating and communicating the embedded value to localvores.

5.2. Consumers' market value

5.2.1. Consumers' profile

A demographic profile of a typical local food consumer, from our sample, was established using the data collected from the respondents who indicated that they currently do purchase local foods. Overall, out of the 1611 surveyed respondents, there are more females (59%) than males (41%), with their ages falling predominately between 35-54 years. The income levels of the respondents varied considerably; the median income range was $60,000 to $90,000 per annum. Further, only 12% of respondents lived on their own; 88% of people lived in households with more than 2 people. Further, the education level of the respondents indicated that the majority of respondents had a college or university degree.

Given that the survey did not define what local food is, it is assumed that the survey respondents understand this concept. Further, there was no differentiation between local and local organic. In terms of local food consumption, 15.9% of respondents indicated they do purchase local organic food, 76.6% do purchase local food (non-organic), and 7.5% did not respond. The objective is to explore whether localvores have refined needs in terms of organic values. With respect to consumer concerns over local food products, 55.9% of respondents did indicate they had concerns over local food products, and 44.1% did not. When asked if those concerns affected their buying patterns, 47% indicated yes, while 53% indicated no. The highest response rate, in terms of weekly purchasing, is that 49.4% of respondents say they purchase between $30 and $50 of local food products. It is also interesting to note that the lowest ranking attribute for all groups (with and without concerns and those who have and have not changed their buying pattern) is the organic attribute. It is evident that organic local is not of importance for the respondents. Maybe because both concepts have overlapping values.

5.2.2. Purchase criteria and preferences

Through factor analysis, it was determined that there is a difference between respondents' answers in terms of their preferred attributes of regular food products versus local food products. The three dimensions considered by consumers when buying local food products in general would include: general product attributes (29.8%), local food dimension (55.3%) and price (7.1%), which collectively explain 62.4% of the total variance. Hence, consumers consider price when deciding to buy local food products. Respondents were asked how much they would be willing to pay for locally produced foods. Of those surveyed, 41.5% indicated they would pay the typical price for regular food products, while 49.1% indicated they would pay 20% premium. Price, however, is not a determinant factor when deciding to buy local food

http://dx.doi.org/10.5772/60565

products. This is consistent with the literature in terms of consumer's willingness to pay a premium for local products. As for the purchasing criteria, two dimensions stand out. These relate to control of the product (54.3%) and food mileage (29.5%) totaling 73.8% of the variance. These dimensions have managerial implications to local food producers when planning to market their food products.

Consumers may express concerns, but seem relatively unknowledgeable and not unified in the nature of the concerns expressed. Frequency analysis reveals that the concerns expressed by the respondents is relatively comparable in all categories (health 29.7%, sanitation 30.5%, production practices 22.5%, antibiotics and hormones 54%). Given that less than one third of the respondents had concerns in all of the respective categories, and also given the somewhat equal distribution, it is noted that there is not a significant concern of one type. It is therefore apparent that the consumers' views on the product attributes (both extrinsic and intrinsic) have more influence on the purchase decisions than the concerns themselves. This is very important as it drives consumers' purchase pattern.

The sample was then divided into two groups, those with concerns over local food products and those without concerns. This helps better understand the differences between both groups' perspective towards general product attributes[2]. Table 2 below illustrates the findings.

Attribute	Concerned	Not Concerned	Significance
Organic	3.8	2.2	**.00***
Taste	4.4	4.3	.05
Price	3.8	3.6	.30
Nutritious and Healthy	4.4	3.6	**.00***
Food Safety	4.7	4.1	**.00***
Fresh	4.7	4.2	**.00***
From Canada	4.4	3.9	**.00***
Local or Regional Brand	4.2	3.9	**.00***
Small Family Farm	3.3	2.9	**.01***

*significance at the.05 level

Table 2. Concerns and Product

In general terms, those concerned with local food products have a higher importance associated with the product attributes, illustrated by the higher means with food safety and freshness being the top scorers for both groups. Also, it is noted that the attributes of price and taste are not statistically different between the two groups (concerned and not concerned); even though price is an important but not a determinant variable.

2 This was done by comparing the means using the 5-point Likert scale; 3 being the median point.

A further comparison was done using the rankings of attributes in comparison to those whose concerns have affected their buying patterns, against those whose have not (cf. Table 3).

Attribute	Purchase pattern affected by concerns	Purchase pattern not affected by concerns	Significance
Organic	3.1	2.2	**.00***
Taste	4.3	4.1	.11
Price	3.7	3.8	.32
Nutritious and Healthy	4.3	3.7	**.00***
Food Safety	4.5	4.4	.11
Product is Fresh	4.5	4.5	.52
From Canada	4.3	4.0	.07
Local or Regional Brand	3.5	3.3	.11
Small Family Farm	3.1	2.9	.22

*significance at the.05 level

Table 3. Impact of Concerns on Buying and Desired Attributes

Overall, those whose concerns have affected their buying behavior, place a higher importance associated with local food attributes. Taste, healthiness, food safety, freshness, and country of origin are important regardless if those attributes are affecting the purchase pattern of respondents. Conversely "organic" and "healthiness" discriminate best between both groups. This is in line with has been said above. Also, it is noted that the attributes of taste, food safety, freshness, and the country of origin (Canada) are not statistically different between the two groups (concerned and not concerned). However, these are very important considerations, given that the means for these attributes are greater than four, on a scale of five points.

5.2.3. Purchasing and consumption patterns

Cross tabulations show that 39% of all respondents indicated that their concerns about local foods affected their buying patterns. The Chi-square test is conclusive ($sig. = 0.000 < 5\%$) which indicates there is an association between local food concerns and local food purchasing. Concerns are therefore determinants of local food consumption and its corresponding purchasing patterns. It is conceivable that food concerns are altering people's decisions to purchase alternative food products. However, ANOVA results indicate that 62.8% of respondents who have concerns do not buy local organic food products. This results is seconded by Chi-square testing, it was determined that concerns are not determinants of current purchases of local organic products. Therefore, contrary to the results of the previous test, in which concern were found to impact sales, these concerns are not related to the decision to buy local organic products. Again, it is conceivable that concerns over local food products in general, are causing people to seek other alternative foods. This is an interesting result as Hamzaoui et al. (2012) show that food concerns drives consumers to choose healthy alternative such as organic foods and to a lower extent local foods. All these results are consistent with the current research and industry data, which indicates that concerns over specific food products have impacted sales.

In comparing the ranking of product attributes, between those who currently indicated that they do buy local food products and those who did not, there is a significant difference between the two groups on many attributes (cf. Table 4): healthy and nutritious, local or regional brand, and small family farm. This is consistent with the literature, in that reduced food mileage often gives consumers the perception that local food has improved health benefits and is fresher. If targeting local food consumers, communicating the benefits of reduced food mileage may be a possible avenue of promotion. Another item of interest is the difference between those who do buy local and those who do not in terms of embeddedness, namely their ranking of the attribute small family farm and local brand. Those who do not intend to buy local foods place a higher importance on these "sustainability" aspects. It is conceivable that the embeddedness theory is correct, in that those who do buy direct from farmers have a greater level of trust and connectedness with the farmer, therefore having fewer requirements for the product to be inspected and tested.

Attribute	Buy Local	Do Not Buy Local	Significance
Organic	3.0	2.8	.08
Taste	4.5	4.2	.12
Price	3.7	3.9	.24
Nutritious and Healthy	4.4	4.1	**.00***
Food Safety	4.3	4.3	.58
Fresh	4.6	4.4	.32
Local or Regional Brand	4.0	3.7	**.00***
Product from Small Family Farm	3.8	3.5	**.00***

*denotes significance at the.05 level

Table 4. Desired Attributes of Local Foods Consumers

Lastly, when ranking the product extrinsic characteristics, it is evident that food packaging can influence a consumer's decision. Consumers in this study indicated a strong importance associated with packaging containing the contact information of farmers as well as the ability to determine the freshness (date of packaging). This finding also would benefit those interested in promoting and retailing their products through grocery stores and supermarkets. Further, respondents indicated that they eat local food products between 1-3 times a week, with shopping occurring at 40.8% at farms' gate (small producers and community farmers), specialty stores (24.9%), community groceries and community chain stores (25.4%), and only

8.9% from farmers' markets. Given that 15.9% of the sample indicated that they do buy local organic products, and only 8.9% do their shopping at the farmers' market, it is questionable whether or not consumers fully understand what local organic is. It may be that when purchasing food products at a local community stores, the consumer might falsely believe they are supporting a local farmer. Furthermore, respondents expressed a willingness to purchase local food products at Supermarkets (22.5%), Local Grocery Stores (93.1%), local community shops (81.8%), local farmer in small quantities (52.4%), in bulk (31.8%) and at local restaurants (60.6%). The high percentages across the board indicate great market potential for farmers to retail their product in all channels, and to strategize better promotion of direct sale of their product.

5.3. Clustering consumers

Based on the previous section results, different combinations of socio-demographic indicators and psychographic variables have been implemented to determine the optimal segmentation strategy. For the purpose of this study, broad segmentation is defined as a segmentation strategy that has non-distinct segments while good segmentation is defined as a segmentation strategy with very distinct segments. The idea is to maximize intra-group homogeneity and intra-group heterogeneity. This allows for more robust profiling, as consumers will behave in the same way when they belong to the same segment and will behave differently if they belong to different segments. Note that homogeneity and heterogeneity are defined with regards to the segmenting variables. For the purpose of having a good measure of inter-group heterogeneity, several ANOVAs were run to make sure that consumers in different segments have different profiles. All tests were conclusive.

The three attributes driving the purchasing behavior and the consumption patter are: organic, food mileage, and healthiness. Consumers were clustered using those 3 variables; all are measured using a 5-point bipolar scale. Results of 2-step cluster analysis show 2 groups of consumers that act in a very distinctive way (cf. Table 5). It is clear that cluster 1, referred to as hardcore local food consumers, is composed of consumers that look for higher and elaborate values than only local. They are looking for local organic food products. Further, they score very high on food mileage and healthiness showing that these attributes are crucial in their purchase decisions. Conversely the regular local food consumers are not looking for any organic dimension rather these consumers look for basic food attribute criteria to make their purchases. These results are in line with the previous results but add a clustering dimension to the results.

		Mean		
Segments	**Percentage**	**Organic**	**Food mileage**	**Healthiness**
Hardcore local food consumers	68%	3.73	4.14	4.43
Regular local food consumers	32%	1.54	3.20	3.67

Table 5. Cluster Analysis Results

http://dx.doi.org/10.5772/60565

6. Discussion and managerial implications

This study has academic and practical implications to both producers and distributors. Findings show that local foods are a by-product and a response to both supply and demand. It is a philosophy of producing and eating food; it is a culture. Consumers' issues and concerns over conventional food products are determinant factors when purchasing local food products. Those who purchase locally grown food might do so because of other factors such as their perceptions of food mileage, packaging, perceived benefits of freshness, and possibly their feeling of contributing to the sustainability of local farms.

[12] argue that distributors are widening their offer and look for more competitive prices. Hence, they relate these trends to the product life cycle. The organic market is driven by conventional marketing strategies and is consistently looking for standardization of the supply. This defeats the intrinsic sustainability objective of such products. This study shows also that most producers/distributors that were interviewed agree on the importance of the production operations and the distribution logistics. There is a clear differentiation between long-medium channels versus short channels. This shows the current divide in the organic food supply and demand sides. Long channels strategies are convenience and price driven. They offer a local value targeted toward a certain consumer profile; these are customers that buy local for health reasons. Conversely, short channels are production method driven. These channels serve consumers having a principle-oriented life style; thus the environment and the support of the local economy are the main drives of this market demand. Price is not an issue here.

For producers/farmers this study has a number of positive outcomes. Intrinsic values such as taste, quality and freshness, nutrition and health, are very important to the consumer both for local food products in general and local organic foods products in particular. Local farmers should target these variables when positioning the product in the region. The reduced food mileage that is achieved when selling local food products from a local farm to a local consumer should be capitalized. There are many benefits to the sustainability of the local farms. With good marketing, local farmers could better promote their product on this basis. Furthermore, given the strong willingness expressed by respondents to purchase local food products if available at all of the various channels of distribution, i.e. local grocers and community groceries, and restaurants, there is great potential for farmers to develop relationships with community businesses.

It is determined that consumers who purchase local food products in general are more price sensitive than those who buy local organic food products. One of the limitations of the study has been that consumers might not *fully* understand where their local food products come from. It is possible that price premiums paid to local food stores might be perceived as a premium on local foods which for the most part is not the product sold at those points of sale. The willingness to pay a premium is consistent with the frequency analysis that reveals that the majority of respondents, who would buy local foods, would pay a premium for this product. As indicated in the U.K. study [25], the interest in buying local often does not develop into the decision to seek local products (between 6-10% of those who say they are interested).

The typical local food consumer in the Ontario and Quebec is consistent with research of [4], that indicates a female, and with a higher-level education. Having profiled this consumer however, it is noted that both the sample and those within the sample who currently purchase local foods are very similar in terms of demographics. It is important to realize that consumers may not fully understand the meaning of locally produced food and demographics alone are not sufficient to explain the purchase behaviour. Given the relationship between concerns and the decision to purchase local food products, local farmers should be pleasantly surprised that price is not the main factor that people consider when deciding to buy local. Future research should be undertaken to assess the effects of different marketing ideas and also to examine if consumers understand the meaning of locally produced food.

To recapitulate, the starting point of the marketing model depicted in Figure 2 starts with the market needs. Depending on the degree of consistency of the need and the knowledge level of the target market, there are two operation sizes: direct (farms gate) or using intermediaries (distributors). Each one of these channels has a value attached to it. The more the consumers know about their needs, the more they will look for an enhanced value capturing all aspects listed in the model. Conversely, if consumers have limited knowledge but are driven by social consciousness (sustainability and helping the local economy), then they will buy from longer channels (specialized, community grocery stores) under the impression that food is local.

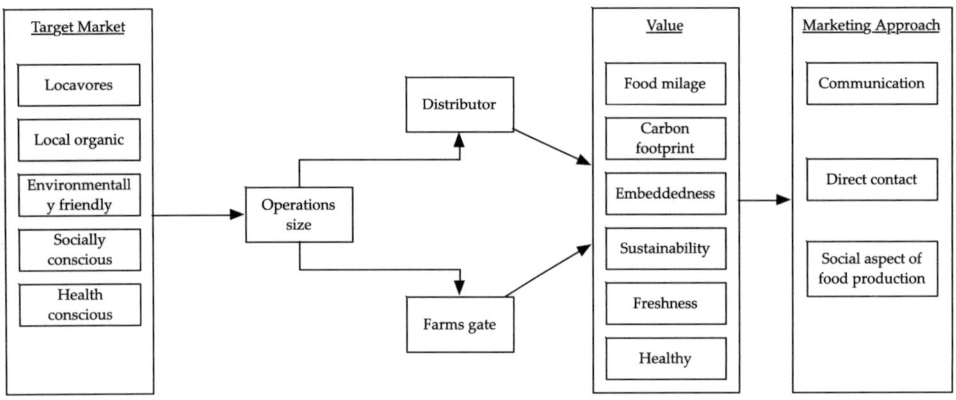

Figure 2. Local Foods Marketing Model

7. Conclusion

Local food research is an area of study with a vast number of possible areas of future research. Local farmers will find value in knowing that market potential does exist for their product, and consumers are expressing an interest in purchasing locally produced food at all channels of distribution. Their motivation to buy local food products is not driven by fear and concerns over food products, but rather by knowing key information on country of origin, nutritional value, freshness and healthiness. This information could be found in the packaging (i.e. date

http://dx.doi.org/10.5772/60565

of packaging, contact information, etc.). In terms of channels of distribution, many more people expressed a willingness to purchase local food products if available at their supermarkets, local grocery stores and specialty stores, while only a very small percentage would consider the farmer's market. This result is consistent with the findings from the study conducted in Ontario [15], which also found a willingness to buy local food products if available in more conventional stores. However, this may defeat the philosophy of buying and eating local. Further, it is important for farmers to note that consumers do express a willingness to pay a premium for their local food products.

Although consistent with other research that has profiled a typical local food consumer, farmers should not solely target the typical demographic profile (well-educated woman, with above average income and family), but should consider the importance of product attributes to all consumers when creating their marketing approach. For example, knowing that a product is locally produced, and promoting it based on its nutrition, health benefits, taste and reduced food mileage, might be a better strategy than just focusing on the typical local foods consumer. Contrary to the existing literature on sustainability, and the concept of embeddedness, this study did not indicate that consumers concerns and or fears changed consumer's decision to buy local. While the study does reveal that concerns have altered the purchasing patterns and behaviours of consumers, these concerns about foods might relate more to the BSE crisis for example, than the fear of the globalized food system. Further, from an academic perspective, the data collected in this study somewhat contradicts the embeddedness theory, given that it has been determined that it was not concerns over the global food system that drive the decision to purchase local food products. Further exploration of the reasoning behind the decision to buy local could be explored in order to determine if social theory, and the desire to purchase sustainable products plays a role in consumers decision-making.

Author details

Mehdi Zahaf*

Address all correspondence to: zahaf@telfer.uottawa.ca

Telfer Scchool of Management, University of Ottawa, Canada

References

[1] Baker, S., Thompson, K., and Engelken, J. (2004). Mapping the Values Driving Organic Food Choice. European Journal of Marketing. 38, 995-1012.

[2] Balderjahn, I. (1988). Personality variables and environmental attitudes as predictors of ecologically responsible consumption patterns. Journal of Business Research. 17 (1), 51-56.

[3] Berger, I., and Corbin, R. (1992). Perceived consumer effectiveness and faith in others as moderators of environmentally responsible behaviors. Journal of Public Policy and Marketing. 11 (2), 79-89.

[4] Conner, D., Colasanti, K., Ross, R. B., and Smalley, S. (2010). Locally grown foods and farmers' markets: consumer attitudes and behavior. Sustainability. 2, 742-756.

[5] Ellen, S.P., Wiener, J., and Cobb-Walgren, C. (1991). The role of perceived consumer effectiveness in motivating environmentally conscious behaviors. Journal of Public Policy and Marketing, 10 (2), 77-101.

[6] Food and Agriculture Organization (2006). Food Security. Retrieved 01 24, 2014, from FAO: http://www.fao.org/forestry/13128-0e6f36f27e0091055bec28ebe830f46b3.pdf

[7] Feagan, R. and Morris, D. (2009). Consumer quest for embeddedness: a case study of the Brantford farmers' market. International Journal of Consumer Studies. 33, 235-243.

[8] Fotopoulos, C., and Krystallis, A. (2002). Purchasing Motives and Profile of Greek Organic Consumer: A Countrywide Survey. British Food Journal. 104, 730-764.

[9] Food Processing Centre (2001). Attracting consumers with locally grown products. Institute of Agriculture and Natual Resources, University of Nebraska – Lincoln.

[10] Future of Food (2015). Retrived March 2015 from: http://www.futureoffood.ox.ac.uk/what-food-system

[11] Hamzaoui-Essoussi, L. and Zahaf, M. (2009). Exploring the Decision Making Process of Canadian Organic Food Consumers: Motivations and Trust Issues. Qualitative Market Research. 12 (4), 443-459.

[12] Hamzaoui-Essoussi, L., Sirieix, L, and Zahaf, M. (2012). What Would Make Consumers Trust Organic Products? A Qualitative Study Based on The Distributors, A Perspective. Proceedings of the ECO-ENA: Economics & ECO-Engineering Associate, Ottawa, Canada, 33-52.

[13] Kassarjian, H., (1977). Content Analysis in Consumer Research. Journal of Consumer Research. 4 (1), 8-18.

[14] Larue, B., West, G., Gendron, C., and Lambert, R. (2004). Consumer Response to Functional Foods Produced by Conventional, Organic, or Genetic Manipulation. Agribusiness. 20, 155-166.

[15] Northwest Link (2009). Retrived July 2009 from: http://www.ontariosoilcrop.org/docs/01NWLApr._09.pdf

[16] O'Hara, S. and Stagl, S. (2001) Global food markets and their local alternatives: A Socio-economic perspective, *Population and Environment*, 22(6), 533-554.

[17] Onozaka, Y., Nurse, G., and McFadden, D.T. (2010). Local Food Consumers: How Motivations and Perceptions Translate to Buying Behaviour. Choices Magazine.

[18] Pirog, R (ed). Ecolabel Value assessment: Consumer and Food Business Perceptions of Local Food. Leopold Center for Sustainable Agriculture and ISU Business Analysis Lab, Ames, Iowa, 2003.

[19] Pirog, R. (2004). Food Miles: A Simple Metaphor to Contrast Local and Global Food Systems. Hunger and Environmental Nutrition, American Dietetic Association.

[20] Pirog, R. and Larson, A. (2007). Consumer perceptions of the safety, health, and environmental impact of various scales and geographic origin of food supply chains. Leopold Centre.

[21] Shamdasni, P., Chon-Lin, G., and Richmond, D. (1993). Exploring Green Consumers in an Oriental Culture: Role of Personal and Marketing Mix Factors. Advances in Consumer Research. 20, 488-493.

[22] Smithers, J., Lamarche, J., and Alun, J. (2008). Unpacking the Terms of Engagement With Local Food at the Farmers,Äô Market: Insights From Ontario. Journal of Rural Studies. 24 (3), 337-350.

[23] Sonnino, R. (2007). The power of place: Embeddedness and local food systems in Italy and the UK, Anthropology of Food. 2.

[24] Tellström, R., Gustafsson, I., and Mossberg, L. (2006). Consuming heritage: The use of local food culture in branding. Place Branding. 2(2), 130-143.

[25] Weatherell, C., Tregear, A., and Allinson, J. (2003). In search of the concerned consumer: UK public perceptions of food, farming and buying local. Journal of Rural Studies. 19, 233-244.

[26] Wier, M., and Calverly, C. (2002). Market Potential for Organic Foods in Europe. British Food Journal. 104, 45-62.

[27] World Food Summit. (1996). World Food Summit. Retrieved 11 6, 2014, from FAO: http://www.fao.org/docrep/003/w3613e/w3613e00.htm

[28] Zanoli, R., and Naspetti, S. (2002). Consumer Motivations in the Purchase of Organic Food: A Means End Approach. British Food Journal. 104, 643-653.

Chapter 2

Biotechnological Production of Oligosaccharides — Applications in the Food Industry

Tathiana Souza Martins Meyer, Ângelo Samir Melim Miguel, Daniel Ernesto Rodríguez Fernández and Gisela Maria Dellamora Ortiz

Additional information is available at the end of the chapter

http://dx.doi.org/10.5772/60934

Abstract

Oligosaccharides are carbohydrates, composed of up to twenty monosaccharides linked by glycosydic bonds, widely used in food and pharmaceutical industries. These compounds can be obtained by extraction from natural sources (milk, vegetables, fruits), and by chemical or biotechnological processes. In the last case, chemical structures and composition of the generated oligosaccharides depend on the type and source of enzymes, and on process conditions, including the initial concentration of substrate. Among the various functions of nondigestible oligosaccharides, one that has attracted attention is its prebiotic potential. The intestinal benefits of prebiotics, such as fructooligosaccharides and inulin as well as their symbiotic association with probiotic bacteria, encompass prevention and treatment of infectious diseases, including viral or bacterial diarrhea, and chronic inflammatory diseases such as ulcerative colitis. Other benefits attributed to prebiotics and probiotics include treatment of inflammatory intestinal and irritable bowel syndrome, prevention of cancer, and modulation of the immune system, mineral absorption and lipid metabolism. Fructooligosaccharides (FOS), galactooligosaccharides (GOS) and chitooligosaccharides (COS) have been widely studied for their prebiotic properties. Moreover, novel oligosaccharides with potential prebiotic activity are currently under investigation. This review will focus mainly on the biotechnological production, health benefits and applications of non-natural oligosaccharides in the food industry.

Keywords: oligosaccharides, biotechnological production, applications, bioactivity

1. Introduction

Consumers all around the world are increasingly aware and concerned about safety and the quality of food. Besides the push towards replacement of chemical additives by those obtained from natural sources, this awareness has led to a rising demand for enrichment of foods with bioactive compounds that have beneficial effects on human health [1]. Therefore, nowadays, a variety of gluten free and products enriched with dietary fiber, or containing probiotics and/ or prebiotic and functional oligosaccharides are available in the market [2].

Oligosaccharides are carbohydrates, composed of up to twenty monosaccharides linked by glycosydic bonds, widely used in food and pharmaceutical industries. These compounds are obtained from natural sources and through chemical or biotechnological processes [3,4].

Among the various functions of non-digestible oligosaccharides, one that has attracted attention is its prebiotic potential. A prebiotic can be defined as "selectively fermented ingredients that allow specific changes, both in the composition and/or activity in the gastro-intestinal microbiota that confers benefits upon host well-being and health" [5]. An oligosaccharide to be regarded as prebiotic must not be hydrolyzed or absorbed in the upper part of the gastrointestinal tract; and must be assimilated selectively by one or by a limited number of beneficial microorganisms in the colon, promoting benefic luminal or systemic effects. To improve colonic function, live microorganisms can be administered in adequate amounts, being known as probiotics; and to be used in food, these organisms must be able to survive passage through the gut; to proliferate and to colonize the digestive tract; and must be safe and effective [6,7].

The intestinal benefits of prebiotics, such as fructooligosaccharides and inulin as well as their symbiotic association with probiotic bacteria, encompass prevention and treatment of infectious diseases, including viral or bacterial diarrhea, and chronic inflammatory diseases such as ulcerative colitis [8]. The mechanisms of action of probiotics against gastrointestinal pathogens consist mainly on competition for nutrients and sites of access, production of antimicrobial metabolites, changes in environmental conditions, and modulation of the immune response of the host. Other benefits attributed to prebiotics and probiotics include treatment of inflammatory intestinal and irritable bowel syndrome, prevention of cancer, and modulation of the immune system, mineral absorption and lipid metabolism [8,9].

Oligosaccharides can be obtained by extraction from natural sources (milk, vegetables, fruits), and by chemical or biotechnological processes [10,11]. Mixtures of oligosaccharides with different degrees of polymerization and glycosidic linkages are usually formed in the enzymatic processes. Chemical structures and composition of these mixtures depend on the type and source of enzymes, and on process conditions, including the initial concentration of substrate [11,12]. Depending on the initial substrate, production of oligosaccharides can involve different steps: hydrolysis of glycosidic bonds giving rise to monomers, followed by generation of disaccharides and other oligomers through the action of transferases [13,14].

http://dx.doi.org/10.5772/60934

2. Fructooligosaccharides

Fructans are carbohydrates in which one or more fructosylfructose links constitute the majority of glycosidic bonds [15]. These carbohydrates can be of the inulin-type with β-(2,1)-D-fructofuranosyl units, found in plants and synthesized by fungi. Additionally, there are the levan-type fructans with β-(6,2)-D-fructofuranosyl units, found in plants and synthesized by bacteria [16].

Levan is a polymer with very high molecular weight that can reach 10^7 Da [17]. In contrast to levan, inulin from chicory consists of a mixture of oligomers and polymers with a degree of polymerization (DP) that varies from two to approximately sixty units (Figure 1; Table 1) [18]. Around 10% of the fructan chains in native chicory inulin have a DP in the range between two and five [5].

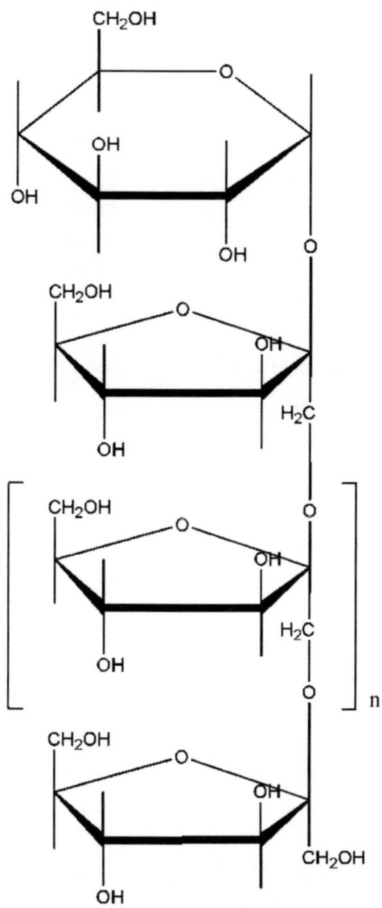

Figure 1. Structure of inulin, a linear fructosyl polymer linked by β-(2,1) bonds (n=3-65), attached to a terminal glucosyl residue by an α-(1,2) bond.

Prebiotics	Chemical structure	Properties	Applicability / Health benefits	Reference
FOS	Fructosyl units linked by β-(2,1) bonds, attached to a terminal glucosyl residue by α-(1,2) bond (Variants: Inulin type β-1,2 and Levan type β-2,6 linkages between fructosyl units in the main chain) DP=3-10.	Soluble fibers; Gel formation; Sugar replacement; Moderate sweetness; Stable (depending on matrix).	Prevention of intestinal infections and extra intestinal infections (e.g. respiratory tract; Inhibition of pathogens, ordering intestinal flora; Regulation of intestinal immune system; Enhancement of immune response; Stimulation of probiotic growth of Lactobacilli and Bifidobacteria species; Optimization of colonic function and metabolism; Production of short chain fatty acids; Increase of mineral absorption; Reduction of food intake and obesity management and control of diabetes type 2; Prevention of cancer.	[8,24,25, 61,67,69, 72,170, 176-186]
Inulin	Mixture of linear fructosyl polymers and oligomers (DP = 3-65) linked by β-(2,1) bonds, attached to a terminal glucosyl residue by α(1—2) bond.	Soluble fiber; Water adsorption; Gel formation; Modifier of viscosity, texture, colour and sensory aspects of food formulations; Replacement for fat and sugar; Low calorimetric value; Moderate sweetness.	Stimulation of probiotic growth; Lowering effect on cholesterol LDL and triglycerides levels; Influence on inflammatory markers and development of gut associated lymphoid tissue (GALT); Regulation of intestinal immune system; Enhancement of immune response; Increase of mineral absorption (Calcium, Iron and Magnesium); Prevention of cancer.	[8,19,67, 176,187-195]
GOS	Mixture of galactopyranosyl oligomers (DP= 3-8) linked mostly by β-(1,4) or β-(1,6) bonds, although low proportions of β (1,2) or β-(1,3) linkages may also be present. Terminal glucosyl residues	Stable in acidic conditions and in higher temperatures; Soluble; Cryoprotector activity; Low ability to crystalyze; Incorporated in various functional foods.	Stimuli of probiotic growth; Reorder intestinal flora; Regulation of intestinal immune system; Reinforcement of intestinal barrier; Inhibition of adhesion of pathogens; Mimic molecular receptors, inhibit microbial adherence;	[8,105, 196-206]

Prebiotics	Chemical structure	Properties	Applicability / Health benefits	Reference
	are linked by β-(1,4) bonds to galactosyl units.		Prevent infections (e.g. *Clostridium difficile* diarrhea); Prevention of cancer; Enhance mineral absorption; Reduce food intake, helping obesity management. Use in diabetic foods, free from carbohydrates that increase the level of postprandial glucose; Use in specialized foods for individuals intolerant to lactose.	
Lactulose	Galactosyl β-(1,4) fructose.	Sweetener, sugar replacement;	Induces growth of *Bifidobacterium* both *in vitro* and *in vivo*; use as laxative in the treatment of constipation; Optimization of colonic function and metabolism, reducing colon pH and ammonia concentration; Increased mineral absorption; Treatment of portal systemic encephalophathy and chronic constipation; Uses in diabetic and dairy foods.	[205, 207-211]
COS	Chitin: β-(1,4) linked N-acetyl-D-glucosamine residues; Chitosan: β-(1,4) linked D-glucosamine polymer. DP=2-8	Antimicrobial activity of chitosan depends on degree of polymerization, amino groups content and degree of acetylation; Chelation of metal trace elements and essential nutrients; Flocculation and adsorption capacity mainly because of the cationic macromolecular structure.	Antimicrobial and antioxidant activity; Use as food preservative; Use as dietary supplements in functional foods; Prebiotic activity; Hypocholesterolemic;	[149, 212, 213]

Prebiotics	Chemical structure	Properties	Applicability / Health benefits	Reference
XOS	Xylose oligomers connected by β-(1,4) linkages (DP=3-6).	Stable in a large range of pH values (2,5-8,0); Thermal stability (up to 100°C); Antioxidant effects; Antifreezing activity; Low cariogenicity; Low calorimetric value; Low glycemic index.	Inhibition of pathogens growth, reordering intestinal flora; Stimulation of probiotic growth; Reinforcement of intestinal barrier; Optimization of colonic function and metabolism; Obesity management, reduction of food intake and weight.	[159, 162, 179, 214-216]
IMO	Glucosyl residues linked to maltose or isomaltose by α-(1,6) glycosidic bonds.	Low sweetness; Low viscosity; Bulking properties; Humectant; Prevention of sucrose crystallization.	Optimization of colonic function and metabolism, reduces nitrogenated products; Increase caecum weight; Antidiabetic effects; Improve lipid metabolism and obesity management.	[2, 217-219]
SOS	Oligomers composed by galactosyl units linked to sucrose by α-(1,6) bonds. Most abundant are raffinose and stachyose.	Stabilizer properties; Cryoprotectant effect.	Prevention of pathogen proliferation.	[2, 156, 213, 220, 221]

FOS: Fructooligosaccharides; GOS: Galactooligosaccharides; COS: Chitooligosaccharides;

XOS: Xylooligosaccharides; IMO: Isomaltooligosaccharides; SOS: Soybean oligosaccharides

Table 1. Structure and biological activity of prebiotics.

Inulin-type fructooligosaccharides are made up of two or more fructosyl moieties linked by β-(2,1) bonds and united at the non-reducing end to a terminal glucose residue by an α-(1,2) glycosidic bond (Table 1) [19]. The term fructooligosaccharides (FOS) is mainly used for fructose oligomers that contain one glucose unit and from two to four fructose units bound together by β-(2,1) glycosidic linkages [20,21]. Nevertheless, oligofructose and FOS may be regarded as synonyms for the mixture of small inulin oligomers with DP<10 [6,22]; while short chain FOS (sc-FOS) are fructose oligomers mainly composed of 1-kestose (GF_2), nystose (GF_3), and ^{1}F-fructofuranosylnystose (GF_4) (Figure 2) [23-25].

Fructans have storage and protective functions in many commonly consumed plants, being a typical part of the diet. Some food sources are richer in high molecular weight fructans, such as inulin, while others have higher levels of sc-FOS [26].

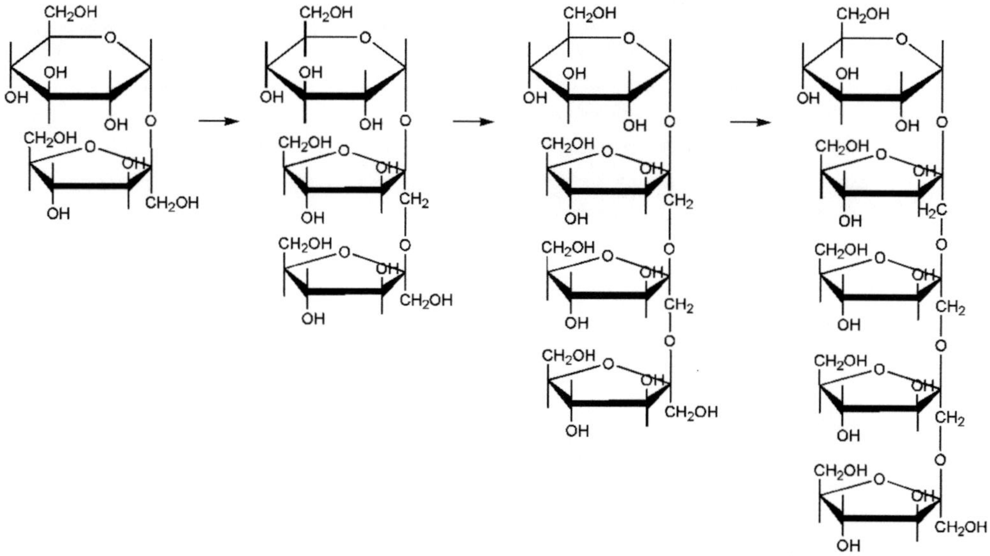

Figure 2. Structures of typical fructooligosaccharides (FOS), derived from sucrose. FOS consist of a glucosyl residue α-(1,2) linked to two or more β-(1,2) fructosyl units. Synthesis of these FOS is catalyzed by fructosyltransferases, requiring a second sucrose molecule as a fructosyl residue donor.

FOS are found in low levels in natural sources such as asparagus, sugar beet, garlic, chicory, onion, Jerusalem artichoke, wheat, honey, banana, barley, tomato, and rye [27-29]. Apart from usually occurring in low concentrations, seasonal conditions also limit their large-scale production from these sources [30].

For this reason, enzymatic processes are used for the industrial production of FOS. One route involves the controlled hydrolysis of long chain fructans (Table 2) [31,32], which results in a large amount of FOS mostly without glucose in their structures. The other route is the synthesis from sucrose, which leads to sc-FOS that contain a molecule of glucose in their structures [11, 33]. The present review will focus on the synthesis of FOS from sucrose.

FOS are produced from sucrose by the action of microbial enzymes with high transfuctosylating activity: β-D-fructosyltransferase (FTase, EC 2.4.1.9) and β-fructofuranosidase (FFase, EC 3.2.1.26) (Table 2) [34]. Since FTase possesses almost only the transfructosylating activity, it is able to cleave the β-1,2 linkage of sucrose, transferring the fructosyl group to an acceptor molecule, with the resulting formation of fructooligosaccharides and release of glucose [35]. This enzyme shows little affinity towards water as an acceptor, therefore the hydrolase activity of FTase is very low [36].

FFase can catalyze both hydrolytic and transfructosylating reactions, nevertheless, transfructosylation only takes place when sucrose concentrations are higher than 500 g L^{-1} [27,34,36-38]. The production of FOS by the action of FFase on sucrose can occur either by reverse hydrolysis or by transfructosylation [36].

Type of prebiotics	Obtention source	Enzyme processing	Microbial producer	Industrial product and manufacturer	References
FOS	Enzymatic reactions: fructosyltransferases using sucrose as a substrate or from inulin using microbial endoinulinases.	Fructosyltransferases or β-fructofuranosidases; Levansucrases; Endoinulinases.	*B. macerans* *Z. mobilis* *L. reutri* *A. niger* *A. japonicus* *A. foetidus* *A. sydowi* *A. pullans* *C. purpurea* *F. oxysporum* *P. citrinum* *P. frequentans* *P. spinulosum* *P. rigulosum* *P. parasitica* *S. brevicaulis* *S. cerevisiae* *K. marxianus*	Neosugar Actilight NutraFlora P-95 - GTC Nutrition Raftilose P95 - Orafti Group	[21,29, 170, 176, 222]
Inulin	Natural product, extraction from plants	Not applicable	Not apllicable	Inulin-S – SigmaAldrich Fibruline - Trades S.A. Fibrex - Danisco Sugar Frutafit CLR DP8, Fruta- fit HD DP10, Frutafit TEX DP5, Inulin TEX – Sensus Inulin GR, HP, HP-gel, HPX, LS, ST, Raftilin ST, Raftilose P95, Raftiline HP - Orafti Group	[189, 194, 222]
GOS	Enzymatic transgalactosylation reactions, using lactose as substrate; Fermentation process.	β-Galactosidases	*Aspergillus sp.* *Bacillus sp.* *B. circulans* *Kluyveromyces sp.* *B. bifidum*	Vivinal GOS Syrup - Bolculo Domo or Friesland Foods Domo	[105,120, 198, 202-205, 223-227]

Type of prebiotics	Obtention source	Enzyme processing	Microbial producer	Industrial product and manufacturer	References
			S. singularis *S. thermophilus* *C. laurentii*	Purimune - GTC Nutrition Oligomate 55NP - Yakult Pharmaceutical Inc. Cup Oligo H-70® Kowa Company BiMuno - Clasado Ltd.	
Lactulose	Thermal-alkaline isomerisation of lactose; Enzymatic transgalactosylation of fructose.	β-Galactosidases β-Glycosidases	*A. oryzae* *S. fragilis* *K. lactis* *P. furiosus* *S. solfataricus*	Sigma Aldrich Discovery Fine Chemicals Solvay	[209, 210, 228-230]
XOS	Enzymatic degradation of xylans	Endo-β-1,4-xylanases, exo-β-1,4-xylosidases, α-glucuronosidases, α-L-arabinofuranosidases, acetylxylan esterases, ferulic acid esterases and p-coumaric acid esterases.	*T. reesei* *T. harzianu* *T. viride* *T. koningii* *T. longibrachiatum* *P. chyrosporium* *G. trabeum* *A. oryzae*	Xylooligo™- Suntory Ltd. YOGHURINA - Suntory Ltd. MARUSHIGE GENKISU - Marushige Ueda Co. L-ONE - Enzamin Laboratory Inc. SUKKIRI KAICHO Lotte Co.	[152, 159, 232]
COS	Enzymatic or chemical depolimerization and deacetylation of chitin and chitosan	Chitosanases and other non-specific enzymes (papain, and lysozyme)	*S. coelicolor* *B. pumilus* *Bacillus sp.* *S. kurssanocii*	Qingdao BZ-Oligo Co, Ltd. BioCHOS. AMSBIO	[213 232-235]
SOS	Directly extracted from soybean	Not applicable	Not applicable	Soya-oligo - The Calpis Food Industry Co.	[152]
IMO	Enzymatic hydrolysis of starch	α-Amylases or pullulanases, β-amylases and α-glucosidases in sequence. Pullulanases	*A. niger* *Bacillus spp.* *B.subtilis* *B. stearothermophilus*	Isomalto-900 - Showa Sangyo	[12, 152, 159, 236]

Type of prebiotics	Obtention source	Enzyme processing	Microbial producer	Industrial product and manufacturer	References
			and T. maritima		
			A. carbonarious		
			L. mesenteroides		

Table 2. Obtention and industrial production of prebiotics.

FOS are produced at industrial scale from concentrated sucrose solutions using fungal transfructosylating enzymes mainly from strains of *Aspergillus niger, Aspergillus oryzae* and *Aureobasidium pullulans* [27,29,30]. Moreover, production of FTase from bacteria (*Lactobacillus*) and yeasts (*Rhodotorula, Candida, Cryptococcus* sp) has been reported [39,40]. The main enzymes used for industrial production of FOS generally give rise to a mixture of molecules with the inulin-type structure, ^{1}F-FOS, whereas those from yeasts usually form levan-type FOS (^{6}F-FOS) or neoFOS (^{6}G-FOS) [41].

The enzymes from *Aureobasidium pullulans* and from *Aspergillus niger* are highly regiospecific in the fructosyl transfer reaction, transferring one fructosyl moiety from sucrose to the 1-OH of the furanoside of another fructose molecule or fructooligosaccharide, with high selectivity [27]. This synthesis is a complex process in which several reactions occur simultaneously, both in parallel and in series, because sc-FOS are also potential substrates of FTase [42].

Catalytic and physicochemical properties of the producing enzymes, as well as production conditions and composition of FOS are different, depending on the microbial strain. For instance, fungal FTases have molecular masses ranging between 180,000 and 600,000, and are homopolymers with two to six monomer units [43].

Fructosyltransferase from *Aureobasidium pullulans* was submitted to preparative scale chromatographic separation on a weak anion-exchanger [42]. The molecular weight of the enzyme determined by size-exclusion chromatography was 570,000. Analysis of the action of FTase on a FOS substrate (Actilight 950P) showed that sucrose was the only donor of fructosyl moiety used in the transfer reaction catalyzed by the enzyme, while the acceptor could be another molecule of fructose or FOS [42].

A transferase isolated and purified from *Aspergillus aculeatus* exhibited pH and temperature optima of 6.0 and 60°C, respectively, remaining stable with no decrease in activity after 5 h under such conditions [44]. The enzyme was monomeric with a molecular mass of 85 kDa. On the other hand, FFase I from *A. pullulans* DSM2404 had a molecular weight of 430,000 [45]. The biocatalyst from *A. aculeatus* showed both transfructosylation and hydrolytic activity, and the transfructosylation ratio increased to 88% at 600 mg mL^{-1} of sucrose [44]. Conditions such as sucrose concentration (400 mg mL^{-1}), temperature (60°C) and pH (5.6) favored synthesis of high levels of GF_3 and GF_4. The major products were GF_2 after 4 h and GF_4 after 8 h of reaction. Prolonged incubation for 16 h resulted in the conversion of GF_4 into GF_2 due to hydrolase activity.

The theoretical yield of FOS from sucrose is 75% if 1-kestose is the only FOS produced [46]. However, production yields of FOS are typically low (55–60%) due to the hydrolytic activity which gives rise to glucose and fructose as reaction byproducts [27] and/or to the fact that glucose acts as an inhibitor of the enzymes, reducing the reaction efficiency [36,47,48]. To improve FOS production yields, glucose oxidase has been used to remove glucose via transformation to gluconic acid [49] and glucose isomerase has been used to interconvert glucose to fructose [46]. Nevertheless, it is necessary to seek for strains among the microbial diversity with high transfructosylating activity, able to produce high yields of oligosaccharides and low yields of monomeric sugars [35].

In addition, the supply of sc-FOS is limited compared to their increasing demand in the food industry, because enzymes such as fructosyltransferases are not widely commercially available [50]. For this reason, the production of FOS is usually carried out in a two-stage process, in which the first stage consists of the microbial production of the enzyme with transfructosylation activity, while the second involves the reaction of the produced enzyme with sucrose (substrate) to generate FOS [29].

A commercial pectinase preparation from *Aspergillus aculeatus*, Pectinase Ultra SP-L, contains FTase [51,52] besides being composed of different pectinolytic and cellulolytic enzymes. The preparation, used in the food industry to reduce the viscosity of fruit juices [42,53], was the only commercially available source of FTase according to [42].

Enzymes from *Aspergillus japonicus*, *Aspergillus aculeatus* (Pectinex Ultra SP-L) and *Aureobasidium pullulans* were used to determine the reaction conditions required to obtain high yields of sc-FOS [34,51,54]. High concentrations of sucrose (600-850 g L^{-1}), pH (4.5–6.5), temperature (50–60°C), reaction time (3–5 h) and high ratios of transferase and hydrolase activities of the enzyme favored transfructosylation over hydrolysis reaction [44,53].

In a recent study, twenty-five commercial enzyme preparations used in the food industry were screened for transfructosylation activity. Three preparations showed high transfructosylation activity from sucrose, high ratios of transferase over hydrolase activity, selectivity for the synthesis of sc-FOS and did not hydrolyze the produced sc-FOS after a 12 h reaction time [55]. Among these enzymes, a cellulolytic enzyme preparation, Rohapect CM, catalyzed the synthesis of sc-FOS with relatively high production yield (63.8%), under cost-effective conditions of temperature (50°C), sucrose concentration (2.103 M) and enzyme concentration (6.6 TU/mL), which could provide a process with potential application at industrial scale [50].

The synthesis of FOS from sucrose is economically advantageous because sucrose is less expensive than inulin; however, the use of enzymes as catalysts for industrial processes is expensive. Furthermore, the recovery of soluble enzymes for reuse is not economically feasible. In contrast, enzyme immobilization usually confers high storage and long-term operational stability, facilitates the recovery and reuse of the biocatalyst, allowing a cost-efficient use of the enzyme in continuous operation, among other advantages [56,57].

In this context, the commercial enzyme preparation from *Aspergillus aculeatus* (Pectinex Ultra SP-L) has been studied for production of FOS in free and immobilized form. Immobilization of the enzyme onto Eupergit C led to retention of enzyme activity for 20 days of batch operation, and both free and immobilized enzyme produced FOS from sucrose with a yield around 57% [58]. Similarly, production of FOS using the enzyme preparation immobilized onto epoxy-activated Sepabeads EC (Sepabeads EC-EP5) reached a yield of 61% after 36 h of reaction [59].

Synthesis of FOS by dried alginate entrapped enzymes (DALGEEs) was recently reported [60]. FTase from *Aspergillus aculeatus*, contained in Pectinex Ultra SP-L, was entrapped in alginate gel beads, which were then submitted to dehydration. The dried alginate biocatalysts were evaluated for the synthesis of FOS from sucrose in a continuous fixed-bed reactor. A 40-fold enhancement of the space-time-yield of the fixed-bed bioreactor was observed when using DALGEEs compared with conventional gel beads. The fixed-bed reactor packed with DALGEEs presented excellent operational stability since the composition of the outlet was nearly constant during at least 700 h, with an average FOS concentration of 275 g/L.

A partially purified β-fructofuranosidase from the commercial enzyme preparation Viscozyme L was covalently immobilized on glutaraldehyde-activated chitosan particles [61]. Thermal stability of the immobilized biocatalyst was around 100-fold higher at 60°C when compared to the free enzyme. The biocatalyst also showed a high operational stability, which allowed its reuse for at least 50 cycles without significant loss of activity. The average yield of FOS production from sucrose was 55%.

An alternative to the enzymatic production of FOS is the use of either free or immobilized whole cells in bioreactors [62]. Production of these oligosaccharides via fermentation processes has the advantage of obviating purification of FOS-producing enzymes from the cell extracts [29,63,64].

An integrated one-stage method for production of FOS via sucrose fermentation by *Aureobasidium pullulans* was developed and optimized with experimental design tools. To maximize production of FOS, temperature and agitation speed were optimized. A production yield of FOS from sucrose of 64% was obtained in 48 h of fermentation under the optimum conditions (32°C and 385 rpm) [62].

Two filamentous fungi, *Cladosporium cladosporioides* and *Penicillium sizovae*, with mycelium-bound transfructosylating activity were recently isolated. *C. cladosporioides* and *P. sizovae* provided maximum FOS yields of 56% and 31%, respectively. *C. cladosporioides* synthesized a mixture of FOS (^{1}F-FOS, ^{6}F-FOS and ^{6}G-FOS, including a non-conventional disaccharide (blastose)) with different glycosidic linkages, which could afford certain benefits regarding their bioactivity [41].

Two food companies in Japan and Korea use different commercial processes for the continuous production of FOS with immobilized cells of *Aspergillus niger* and *Aureobasidium pullulans*, respectively, both entrapped in calcium alginate gel [27,63]. Calcium alginate has also been employed to immobilize mycelia of *A. japonicus* aiming to establish FOS-producing processes [65,66].

Immobilization of whole cells of *Aspergillus japonicus* ATCC 20236 onto different lignocellulosic materials was also undertaken to produce fructooligosaccharides. Cells immobilized in the different support materials showed FOS production and FFase activity ranging from 128.35 to 138.73 g/L and from 26.83 to 44.81 U/mL, respectively. Corncobs were the best support for immobilization, providing the highest results of microorganism immobilization, FOS and FFase production. In addition, use of immobilized cells led to higher FOS productivity and yield, as well as higher transfructosylation over hydrolysis ratio of FFase than free cells [64].

Several important health benefits are associated with the consumption of FOS as food ingredients. These include modulation of colonic microflora; improvement of the gastrointestinal physiology; activation of the immune system; enhancement of the bioavailability of minerals; reduction of the levels of serum cholesterol, triglycerides and phospholipids; and prevention of colonic carcinogenesis [34,44,67,68].

Among the different FOS, 1-kestose is considered to have better therapeutic properties than those with higher degree of polymerization [69]. The chain length is an important factor influencing the physiological effect of the oligomer in the host [69] and fermentation by bifidobacteria and lactobacilli species [70].

In this context, fermentation of oligosaccharides was evaluated using pure FOS mixtures containing three FOS species (GF_2, GF_3 and GF_4). Only two oligosaccharides (GF_2 and GF_3) were consumed by *Lactobacillus* strains. Moreover, none of the investigated strains metabolized the GF_4 species, suggesting an intracellular metabolism after the FOS transport [70]. This transfer apparently involves an ATP-dependent transport system with specificity for a limited scope of substrates [71].

Moreover, β-fructofuranosidase activity enables bifidobacteria to degrade FOS. Nevertheless, this property is strain-dependent. Some strains consume both fructose and oligofructose, with different preferences and degradation rates [72].

FOS can be used as calorie-free and non-carcinogenic sweeteners. 1-Kestose has enhanced sweetening power compared to other sc-FOS, and 1-kestose-rich sc-FOS syrups can be used as sugar for diabetics [27,73].

Other types of FOS, such as the levan-type and the neo-FOS, have very promising properties; however, they are not yet commercially available [53,74,75].

3. Galactooligosaccharides

Lactose is a disaccharide formed by the condensation of glucose and galactose molecules, and is the most important component of mammalian milk, present in a concentration range from 2.0% to 10%. Lactose can be obtained at industrial scale from whey during cheese production, with dry weight around 80-85%, using crystallization techniques [76-78]. In the past, whey was considered a waste, although, nowadays, it is used to produce whey powders products, improving economic and environmental aspects of the by-products [79].

Lactose presents a great importance for food and pharmaceutical industries, being used in various food products such as chocolate, confectionary and other processed products, as well as carrier of medicines in dry powder inhalation preparations, excipient of tablets [80]. In humans, lactose can cause abdominal discomfort due to its maldigestion, which reaches approximately 70% of the world's adult population [81]. β-Galactosidase (β-D-galactoside galactohydrolase, E.C. 3.2.1.23) plays an important role in human health because it is able to catalyze the hydrolysis of lactose in glucose and galactose, and because of that, it is often referred to as lactase. In addition, the transglycosylation reaction can also occur, in which galactooligosaccharides (GOS) are produced, and their structures can differ in regiochemistry of glycosidic linkage and degree of polymerization (Figures 3-5; Tables 1 and 2) [82,83].

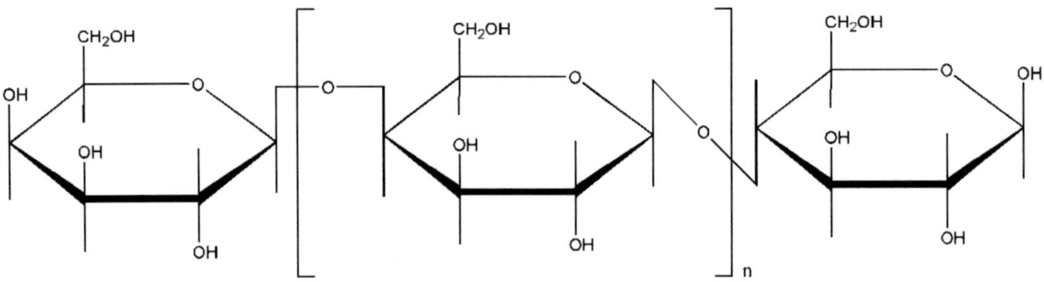

Figure 3. Structure of a galactooligosaccharide (GOS) derived from lactose, a β-(1,4) linked galactosyl oligomer (n=1-4), attached to a terminal glucosyl residue by a β-(1,4) bond. GOS are synthesized by the reverse action of β-galactosidases on lactose in higher concentrations.

Despite the fact that enzymes such as β-glycosidases and β-glucosidases, which also hydrolyze carbohydrates, are able to catalyze transglycosylation reactions, β-galactosidase is the most used enzyme in dairy industry to produce GOS. β-galactosidases from *Kluyveromyces* sp. and *Aspergillus* sp. are the most used in industry because products from those microorganisms are considered as GRAS [84].

Galactooligosaccharides can be defined as a mixture of substances produced from lactose, with two to eight saccharide units, in which one of the units is a terminal glucose and the remaining units are galactose and disaccharides comprising two units of galactose [85]. Several of these GOS are recognized as prebiotics, because they are non-digestible saccharides and can be used selectively by bifidobacteria and lactobacilli in human intestine, and thus improve host health [86, 87].

Conversion of lactose into GOS is catalyzed by β-galactosidases in a kinetically controlled reaction that involves competition between hydrolysis and transgalactosylation. The thermodynamically favored hydrolysis of lactose, which generates D-galactose and D-glucose, competes with the transferase activity that produces a complex mixture of galactose-based di- and oligosaccharides. Transgalactosylation involves direct galactosyl transfer (intramolecular reaction) to D-glucose yielding regio-isomers of lactose, and indirect transgalactosylation (intermolecular) giving rise to disaccharides, trisaccharides, and tetrasaccharides, and eventually longer GOS. The interaction in the active site of the enzyme differs with the acceptor. When the acceptor is water, glucose and galactose are formed; whereas if the acceptor is a sugar, reaction results in GOS [86, 87].

http://dx.doi.org/10.5772/60934

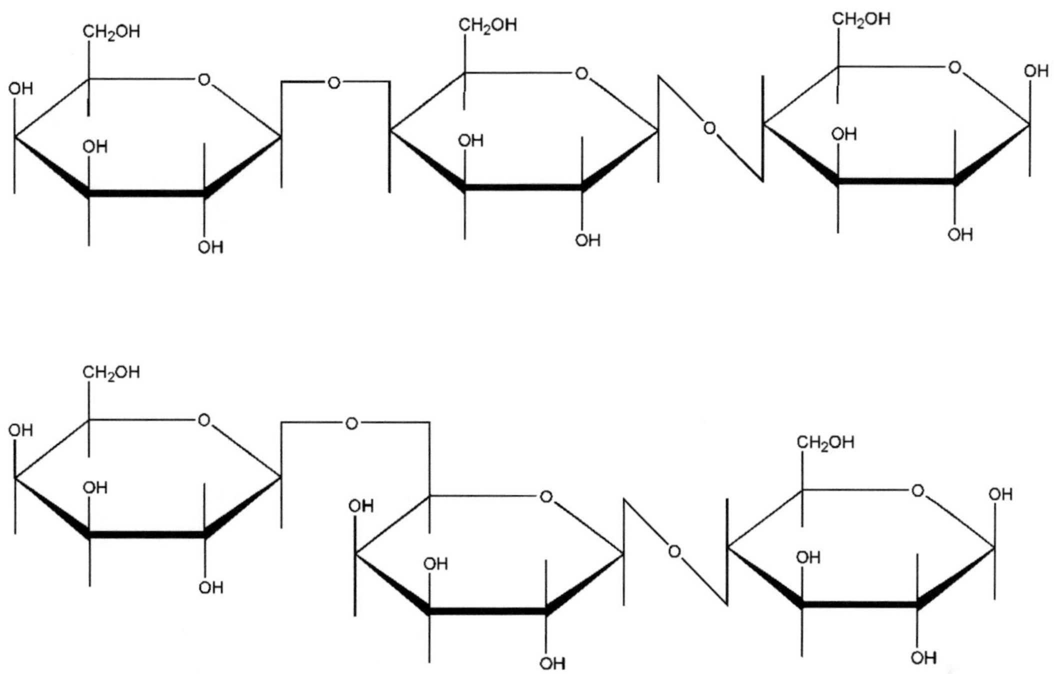

Figure 4. Examples of structures of galactooligosaccharides: 4-galactosyl lactose (top) and 6-galactosyl lactose (bottom) are represented, showing usual regiochemistry differences in galactosyl linkages.

Therefore, high lactose concentrations and low water contents are favorable for GOS synthesis, being the initial lactose concentrations the most important factor, independently of the enzyme source. In general, higher lactose concentrations than 30% are necessary to favor synthesis over hydrolysis [87]. However, at the same lactose concentrations, different yields of GOS can be obtained, because β-galactosidases from different sources, with different structures and/or mechanisms, exhibit different selectivity for water and saccharides. Moreover, GOS yields depend on process conditions, such as temperature, reaction time, pH and enzyme/substrate ratio [88]. However, GOS production can be affected by glucose and/or galactose that are recognized as inhibitors of hydrolysis for many β-galactosidases [89,90].

The reaction time and initial concentration of lactose are considerably important to favor GOS production, since they are simultaneously synthesized and hydrolyzed by β-galactosidase, being regulated by the kinetics of synthesis and hydrolysis. Additionally, lactose concentration can increase formation of GOS due to increased availability of galactosyl and decreased availability of water [82,91]. Additionally, reverse micelle systems, in which the enzyme is entrapped in an aqueous micelle surrounded by organic solvent, provide decrease of the thermodynamic activity of water [92,93]. Chen et al. 2003 [93] reported that the transgalactosylation capability of low concentrations of β-galactosidase and lactose, operating in reverse micelles system, was similar to high concentrations of enzyme and substrate in an aqueous system. Authors also showed that GOS production decreases with the increase in water content.

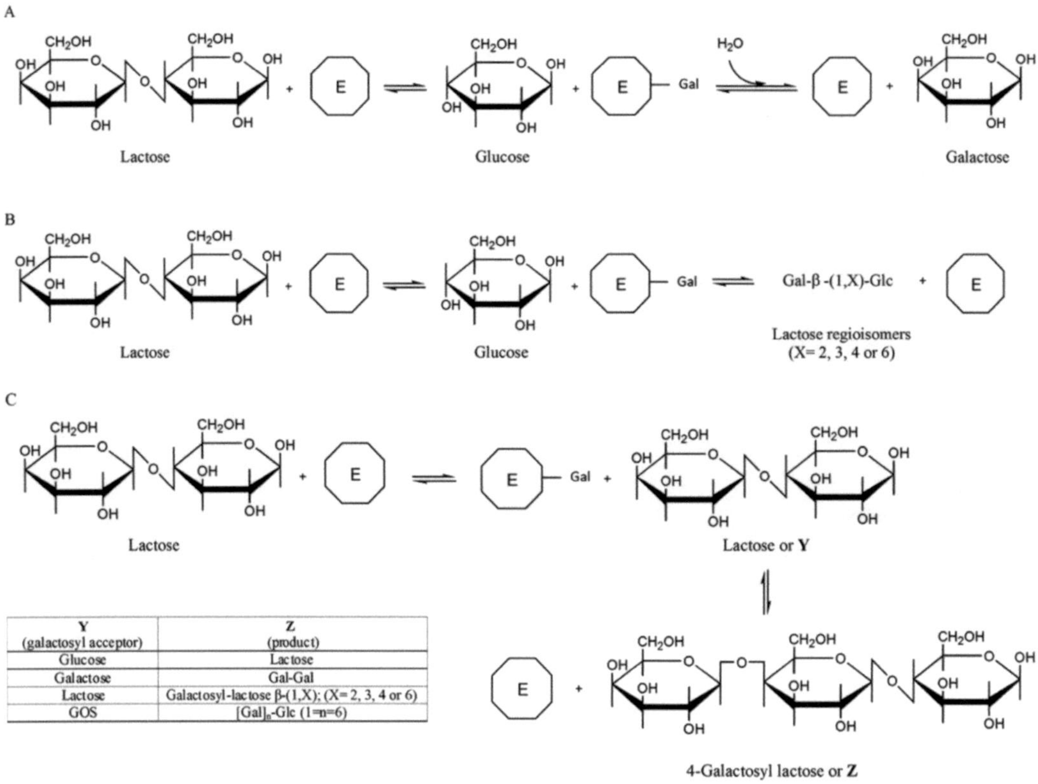

Y (galactosyl acceptor)	Z (product)
Glucose	Lactose
Galactose	Gal-Gal
Lactose	Galactosyl-lactose β-(1,X); (X= 2, 3, 4 or 6)
GOS	[Gal]$_n$-Glc (1=n=6)

Figure 5. Enzymatic synthesis of GOS by transgalactosylation reactions. Transgalactosylation is the transfer of the galactosyl residue, after the cleavage of lactose, to an acceptor molecule containing a hydroxyl group. When the acceptor is water (A), a galactose is formed by lactose hydrolysis, whereas if the acceptor is a sugar, a disaccharide or a GOS may be formed. In intramolecular transgalactosylation (B), galactosyl donor and acceptor are the same (glucose), only linkage position changes. In intermolecular transgalactosylation (C), there is an enzymatic transfer to another nucleophilic acceptor (Y), which can be all the sugars present in the reaction media, resulting in GOS mixtures.

Production of GOS can be improved increasing the reaction temperature. Lactose has relatively low solubility at room temperature, which increases with increasing temperature. Therefore high temperatures are desirable since they allow the increase of lactose concentration [94,95]. Besides the possibility to increase the solubility of subtracts and products, high temperature is advantageous due to the reduced risk of microbial contamination, lower viscosity and improved transfer rates [96]. However, this is not a general rule, Boon et al. (1998) [97] reported that the increase of initial lactose concentration achieved at high temperature does not influence GOS yield using β-galactosidase from *Pyrococcus furiosus*. Another problem of carrying out GOS synthesis at high temperature is the occurrence of Maillard reaction and enzyme inactivation. Bruins et al. (2003) [95] noted that in addition to enzyme inactivation with the increase of temperature (80°C or above), Maillard reactions almost doubled the rate of enzyme inactivation. Therefore, the development of new thermostable enzymes, through recombinant DNA technology, has been undertaken in order to improve the GOS yield

[98-102]. Hansson et al (2001) [103] verified an increase of GOS yield due to an increase of transgalactosylation/hydrolysis ratio by changing a phenylalanine residue to tyrosine in β-glucosidase from *Pyrococcus furiosus*, using site directed mutagenesis.

Another strategy to decrease water activity, and carry out catalysis with both high lactose concentration and temperature, demonstrated by Maugard *et al.* (2003) [104], is the use of microwave irradiation. GOS was produced using immobilized β-galactosidase from *Kluyveromyces lactis* along with organic solvents. In these conditions the selectivity for GOS synthesis was increased 217-fold, compared to a reaction carried out under conventional heating.

Similarly to temperature, pH value can affect the GOS yield, possibly through the control of synthesis and degradation [105] According to Huber *et al.* (1976) [106], that studied β-galactosidase from *Escherichia coli* K-12, higher pH values than 7.8 increased transgalactosylation/hydrolysis ratio, which decreased at lower pH values than 6.0. In contrast, Hsu *et al.* (2006) [107] observed that β-galactosidase from *Bifidobacterium longum* CCRC 15708 exhibits its maximum activity at pH 7.0. This enzyme was stable between pH 6.5-7.0, and after three hours in these conditions, 20% of its activity was lost.

In general, oligosaccharides, including galactooligosaccharides, are produced using sucrose or starch, whey, among other substrates with high quality and low cost. The process designed to convert raw material into oligosaccharides must be inexpensive and focused on increasing the productivity and stability of enzymes. In this context, immobilization of biocatalysts can reduce the process costs due to some advantages; such as possibility to reuse the biocatalyst, applying a series of batchwise or continuous reactions; the biocatalyst can exhibit more stability than the native counterpart; besides this, immobilization can reduce costs of downstream, since separation of the biocatalyst from the product can be minimized [108-110]. Recently, several authors have employed immobilized β-galactosidase to produce GOS, applying different strategies with promising results [111-114]. Urrutia et al (2013) [111] immobilized *Bacillus circulans* β-galactosidase in glyoxyl agarose. The enzyme did not lose the synthetic capacity, and retained 92% of its activity along 10 reaction batches, producing 1956 g GOS/g protein at the end of 10 batches. Palai et al (2014) [112] immobilized β-galactosidase in hydrophobic polyvinylidene fluoride and the reaction for GOS production was carried out with partial recirculation loop. Both GOS concentration and selectivity for GOS production increased with increasing initial lactose concentrations, with maximum GOS production of 30% at 50°C, and feed flow rate of 0.5 mL/min. A novel economic and efficient method to produce GOS through cellulose-binding fusion β-galactosidase was developed by Lu et al (2012) [113]. A fusion protein, formed by β-galactosidase from *Lactobacillus bulgaricus* L3 and a cellulose binding domain were employed for immobilization by adsorption onto microcrystalline cellulose. The immobilization was conducted with efficiency of 61% and the maximum GOS yield was 49% (w/w). Moreover, enzymatic activity of 85% and yield over 40% (w/w) were maintained after twenty batches. Warmerdam *et al.* (2014) [114] carried out GOS production in a packed-bed reactor using commercial β-galactosidase (Biolacta N5) immobilized on Eupergit C250L. GOS productivity was six-fold higher in one run in the packed-bed reactor than observed in one run in a batch reactor.

Smart polymers have been studied to develop GOS production processes. Poly-N-isopropyl acrylamide is a thermo-responsive poly-N-isopropyl acrylamide (PNIPAAm), which presents good solubility in water and distinct phase transition at its lower critical solution temperature (LCST). It is applied in different areas, such as medicine, biotechnology, and engineering [115,116]. Based on these advantages, Palai et al (2014) [117] developed a useful bioconjugate between PNIPAAm and β-galactosidase. The constructed PNIPAAm-β-galactosidase (PNbG) can be used in catalysis and, after that; it can be easily separated from the solution by heating at a temperature above its LCST. Further on, Palai et al (2015) [118] continued the GOS production research using this bioconjugate. A maximum GOS yield of 35 % was obtained at pH 6 and 40°C. An increase in GOS yield was observed when the temperature was risen from 30 to 40°C. At 45°C or above, after prolonged time, enzyme deactivation occurred. Moreover, bioconjugates could be reutilized at least ten times; and the separation was done by simple decantation after addition of 0.05 M NaCl and heating at 40°C.

The use of resting or living cells for GOS production appears to be interesting due to its low cost when compared to the use of purified enzyme. Despite the complexity of biocatalysis processes involving whole cells, glucose and galactose can be consumed by them. The consumption of the monosaccharides is interesting because their presence in foods is undesirable, since they do not exhibit prebiotic effect, increase caloric value of food, and can inhibit the activity of certain β-galactosidases [119].

Nevertheless, the use of whole cells can be exploited in order to selectively improve GOS production [120]. Beta-galactosidase form *Aspergillus oryzae* was used to produce GOS from lactose, followed by fermentation with *Kluyveromyces marxianus* cells, that consumed mono and disaccharides. GOS with 95% purity containing mostly tri- and tetrasaccharides were obtained [120]. Association of β-galactosidase and cells can be applied to develop GOS enriched food products. During yogurt manufacturing, GOS was produced by addition of a commercial β-galactosidase, since starter and probiotic culture were not able to provide it. Thus, this yogurt with low lactose content can be useful for lactose intolerant people. Moreover, GOS was stable during storage, probably because it was not metabolized by microbial culture and enzyme was inactivated by yogurt pH [121].

Products containing GOS were launched for the first time in Japan in the 1980s. Due to their various and important health benefits, applications of GOS gradually increased worldwide. These oligosaccharides can be found in diverse products such as yogurt, bakery products, beverages, snack bars among others [122]. GOS are able to stimulate the growth of bifidobacteria and lactobacilli in the lumen despite other members of the microbiota that were considered potentially harmful. These oligosaccharides can prevent bacterial adherence due to their properties of mimicking host cell receptors in which bacterial adhesion occurs [123]. GOS can hinder the development of colon cancer, effect which can be attributed to their capacity of delaying fermentation processes, and reducing the activity of genotoxic bacterial enzymes associated with this disease [124]. Mineral absorption can be stimulated by GOS administration, and their effect on calcium absorption was verified. GOS can be used to alleviate constipation, which is relatively common in elderly people and pregnant women. It occurs due to increased bacterial growth and fecal weight; besides this, short fatty acids stimulate

http://dx.doi.org/10.5772/60934

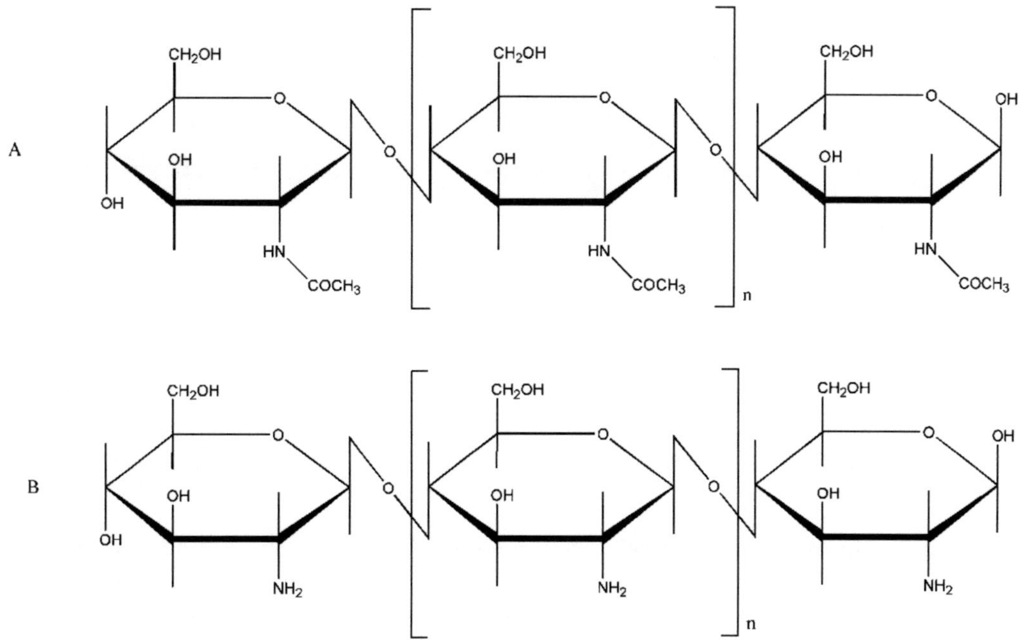

Figure 6. Structures of chitooligosaccharides (n= 3-7), A – Chitin (β-1,4-linked N-acetyl-D-glucosamine residues); B – Chitosan (β-1,4-linked D-glucosamine polymer).

intestinal peristalsis and increase osmotic pressure of fecal weight. Moreover, GOS have been reported as indirectly acting on mucosal and systemic immune activity, and also as having protective effects against allergic manifestations [125].

4. Chitooligosaccharides

In the last years, studies of production and application of chitooligosaccharides (COS) have increased due to their biodegradability, biorenewability, biocompatibility, physiological inertness and hydrophilicity, properties that serve as a basis for the use of COS as functional food or to preserve food from degradation.

Chitin is one of the most abundant natural compounds on earth and its production is mainly based on the extraction from marine species (shrimps, crabs, lobsters, krills, etc.) [126]. Chitin is a copolymer of N-acetyl-D-glucosamine and D-glucosamine units linked by β-(1,4) glycosidic bonds, where N-acetyl-D-glucosamine units are predominant in the polymeric chain as shown in Figure 6A [127]. Chitin obtained from natural sources has a complex composition, containing several minerals, proteins, lipids, pigments and other compounds. Chitosan, an important derivative from chitin, is the deacetylated form of chitin, where N-acetyl groups are removed by chemical methods (Figure 6B).

A considerable amount of residues from processing of fish and crustaceans, rich in chitin and chitosan, are considered hazardous wastes and at the same time have high potential commercial value as raw material [128]. It is possible to obtain chitooligosaccharides from those residues, after prior demineralization and deproteinization by acid and alkali treatments [129].

Chitooligosaccharides are produced by chemical methods or by enzymatic methods from chitosan, produced by alkaline N-deacetylation. At industrial scale, the chemical route is used to produce chitooligosaccharides; however, this methodology presents several disadvantages such as high cost, low yield due to indiscriminate breaks of the polymer chain, production of toxic compounds due to modification on the chitin structure, as well as, corrosion and environmental hazards [130].

The enzymatic process is an attractive solution to overcome the above-mentioned disadvantages, due to their specific action on the substrate, despite the economic costs. Enzymatic hydrolysis of chitin or chitosan involves several enzymes: chitinase, chitosanase, lysozyme and cellulase [131]

According to Mourya *et al.* (2014) [132], various specific enzymes as chitosanases, chitinases and other nonspecific enzymes can hydrolyze chitin and chitosan. Action of chitinases and chitosanases are related to the degree of acetylation of the biopolymers. A novel flow chart for COS production from chitin employing chitinases and chitosanases has been reported (Figure 7) [130].

Chitnases are chitinolytic enzymes hydrolyzing the glycoside bonds between the sugars, which have the unique capacity to hydrolyze the GlcNAc-GlcNAc (2-acetamido-2-deoxy-β-D-glucose) links. Pre-treatment with acid solution is necessary to break down the crystalline structure of chitin and increase the availability of substrate to the action of enzymes. Chitosanases are enzymes that hydrolyze chitosan, classified according to the substrate specificity towards chitosan, which act specifically on the deacetylated (D–D) bonds [133].

In recent years, many scientific papers reported the application of chitinolytic enzymes, from different microorganisms, for the hydrolysis of chitin and chitosan. Enzymes for hydrolysis can be free or immobilized in non-toxic and inert supports.

Fernandes de Assis *et al.* 2010 [134] reported that COS yields of 54% were obtained after 10 minutes of hydrolysis reaction. Initial concentration of chitosan was 1% and the final oligomers concentration was 5.43 mg/mL. Production yields decreased when hydrolysis reaction time exceeded 10 minutes.

Gao *et al.* 2012 [135] determined that the optimal enzyme/chitosan ratio was 7.3 U/mg chitosan at 55°C to produce COS from chitosan employing chitinases from *Bacillus cereus*, achieving a hydrolysis yield of 76%. The yields of COS (GlcN)2, (GlcN)3 and (GlcN)4 were 13.2; 32.6 and 30.2%, respectively.

Ming *et al.* 2006 [136] producing chitooligosaccharides, reported pH range 4.5-6 as the optimal for chitinase activity, reaching 20 g/L of chitooligossacharides from an initial concentration of 50g/L of chitosan, which means a system with 40% of yield in the conversion of chitosan into chitooligosaccharides. Also, employing free and immobilized chitosanase from *Bacillus*

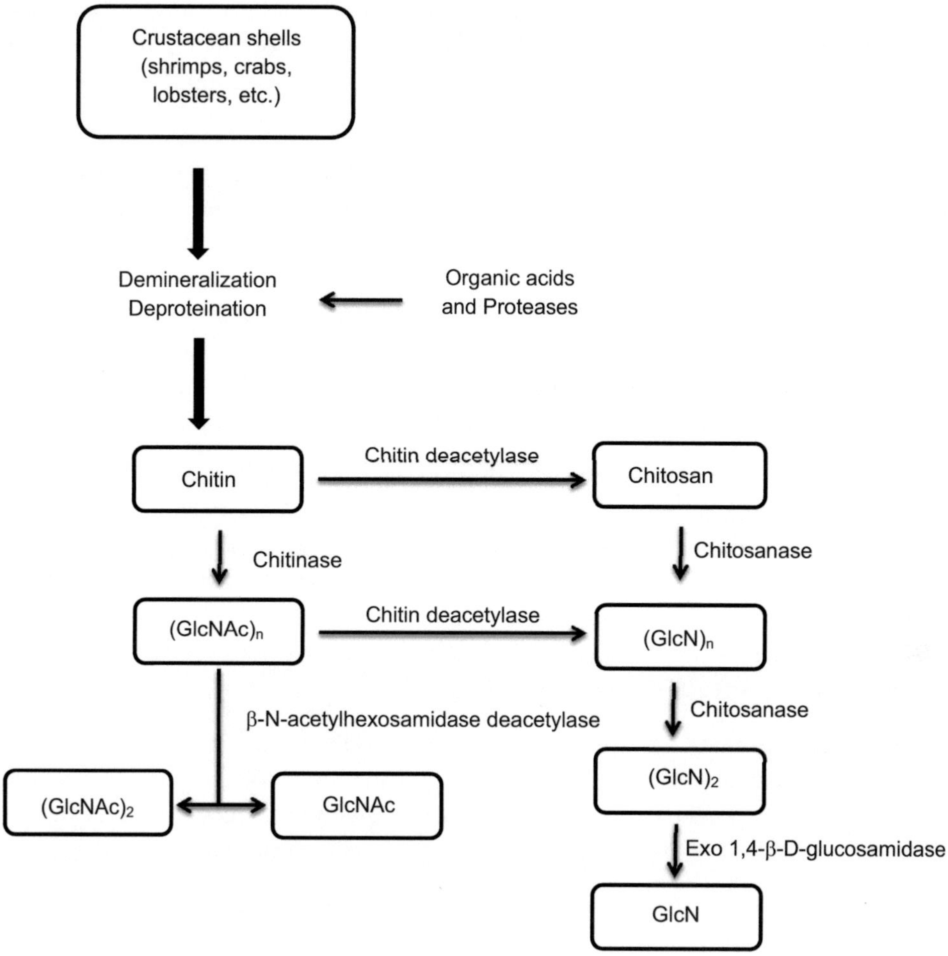

Figure 7. Global flow chart for production of COS by enzymatic hydrolysis of chitin and chitosan (adapted from Jung and Park 2014 [130]).

pumilus, Kuriowa *et al.* 2009 [137], produced chitooligosaccharides in batch and continuous systems. In a system with free enzyme at batch conditions a concentration of 2.8 g/L was achieved, from an initial concentration of 5g/L after 40 minutes of treatment. Another system used was a membrane reactor with cutoff 2000Da. Enzyme concentration of 940 U/L, 40 minutes of residence time and 35°C were reported as the optimal conditions to attain 2.6 g/L of chitooligosaccharides. The membrane bioreactor with the free enzyme was able to maintain a constant rate of chitooligosaccharides production for 96 hours, after that time concentration decreased due to inactivation of enzymes. In order to extend the period of operation, the use of immobilized enzymes was evaluated in the membrane bioreactor. The maximum total concentration of chitooligosaccharides was 2.3 g/L with 620 U/L of immobilized chitinase during 1 month, however it is important to point out that yield was 46%, lower when compared to free enzyme tests.

COS can be applied as food preservatives due to their antimicrobial activity and as functional food, mainly in prebiotics and to help the absorption of important minerals, as calcium. Antimicrobial activity of COS depends on the degree of polymerization (DP) and the degree of deacetylation (DD) as summarized in Table 1.

Inhibitory effects of COS were tested on both Gram (-) and Gram (+) bacteria, including *Escherichia coli, Pseudomonas fluorescens, Salmonella typhimurium, Vibrio parahaemolyticus, Listeria monocytogenes, Bacillus megaterium, Bacillus cereus, Staphylococcus aureus, Lactobacillus plantarum, Lactobacillus brevis* and *Lactobacillus bulgaricus* [138]. Solutions containing 1% (w/v) COS with different molecular weights inhibited bacterial growth by 1-5 log cycles. For Gram (-) bacteria the antimicrobial activity was inversely proportional to the molecular size of oligomers, which means higher antibacterial activity was found with lower molecular weight of oligomers (1 kDa). This phenomenon was not observed for Gram (+) bacteria.

The proposed mechanism of antibacterial activity for COS with DP>12 was cellular lysis [139]. This would be due to the cationic charges of COS that could link to the negative charges present in the cell walls, leading to the formation of large bacterial clusters, which might block the nutrition transport across the bacterial cell and result in death of the bacteria. Highly deacetylated COS were shown to be more effective at inhibiting the growth of *Staphylococcus aureus, Escherichia coli, Pseudomonas aeruginosa, Streptococcus fecalis* and *Samonella typhimurium* than COS with low degree of deacetylation [140].

It has been suggested that COS are able to pass through the bacterial cell wall and be incorporated in the cytoplasm of Gram (+) bacteria [141]. Those low molecular weight compounds can have importance in gene expression related to regulation of stress, autolysis and energy metabolism.

Chitooligosaccharides with DP 4 were demonstrated to have higher antimicrobial effect on four bacteria species (*Escherichia coli, Staphylococcus aureus, Streptococcus lactis, Bacillus subtilis*) and six fungi (*Saccharomyces cerevisiae, Rhodotorula bacarum, Mucor circinelloides, Rhizopus apiculatus, Penicillium charlesii, Aspergillus niger*) [142]. At the same time, degrees of deacetylation over 90% were shown to be more efficient in the inhibition of microbial growth. In addition, chitooligosaccharides with low molecular weight were able to cross the cell wall and interact with DNA in the cytoplasm suppressing the growth of microorganisms. Highly deacetylated COS have many free amines, which can bond to negatively charged residues at the cell wall, leading to formation of aggregates of microorganisms. Those aggregates precipitate, resulting in death of the microorganism.

Chitooligosaccharides can be employed as preservatives due to their antioxidative properties. Antioxidant activity of chitooligosaccharides depends on their degree of deacetylation and molecular weights [143]. It was shown that 90% deacetylated medium molecular weight COS have the highest free radical scavenging activity for DPPH, hydroxyl, superoxide and carbon centered radicals [144]. Antioxidant properties are closely related to the amino and hydroxyl groups, which can react with unstable free radicals to form stable macromolecule radicals [145,146].

According to Halden *et al.* 2013 [147] COS could be applied as feed additives or hypocholesterolemic agents. Based on their study, hypercolesterol concentration in blood is directly related to the generation of reactive oxygen species. Thus, chitooligosaccharides can be used to scavenge the free radicals on the body, triggering the enhanced synthesis of catalase and superoxide dismutase and decreasing lipid peroxidation.

COS were conjugated with phenolic acid (PAC-COS) to improve the antioxidant properties of the oligosaccharides in the presence of reactive oxygen species (2,2-diphenyl-1-picrylhydrazyl (DPPH), hydroxyl (OH) and nitric oxide (NO)) [148]. The increase on the antioxidant activity is associated to the structure of phenolic acids and the substitutions on the aromatic ring of the side chain.

Chitooligosacharides can be considered as prebiotics because they are non-digestible food ingredients with beneficial effects on probiotic bacteria (*Lactobacillus* and *Bifidobacterium*) present in the gastrointestinal tract [5]. In fact, prebiotic activities of COS preparations (0.1 to 0.5%) with varying degree of polymerization (2 to 8) were reported [149]. Assays were conducted with three strains of probiotic bacteria, *Bifidobacterium bifidum KCTC 3440, Bifidobacterium infantis KCTC 3249* and *Lactobacillus casei KCTC 3109*.

However, an opposite effect was shown on the population of *Lactobacillus* and *Bifidobacterium* when chitooligosaccharides were tested as prebiotic agents in healthy rats [150]. Chitooligosaccharides have been demonstrated to have a weaker prebiotic effect over *Lactobacillus* and *Bifidobacterium* when compared with other oligosaccharides as fructooligosaccharides, mannanoligosaccharides; and galactooligosaccharides [151].

Chitooligosaccharides from marine species, mainly shrimps and crabs, are produced and commercialized by several companies (Table 2), such as:

- Qingdao BZ-Oligo Co., Ltd: Monomers of chitosan oligosaccharides are obtained by enzymatic hydrolysis, chemical derivatization and column chromatography. The degree of polymerization is from 2 to 10.
- BioCHOS: Preparation of chitooligosaccharides (CHOS) made by controlled enzymatic degradation of chitosan.
- AMSBIO: Preparation of a series of chitosan-oligosaccharides from dimer to hexamer by hydrolysis of chitosan from crab shells. All oligomers are chromatographically pure, not less than 98%, confirmed by high performance liquid chromatography.

5. Novel oligosaccharides

Typical oligosaccharides like FOS and GOS in particular have been widely studied for their prebiotic effects. However, a number of other non-digestible oligosaccharides (NDOs), to which less rigorous study has been so far applied, have at least indications of prebiotic potential. Those with the most accumulated evidence to date are isomalto-oligosaccharides (IMO), soybean oligosaccharides (SOS), xylo-oligosaccharides (XOS) and lactosucrose.

Together with FOS, GOS, and lactulose, all of these oligosaccharides are recognized in the Japanese functional food regulation system as ingredients with beneficial health effects [152].

A great interest resides on the identification, evaluation and commercialization of new products with improved functional properties and benefic health effects such as higher ability to modulate microbiota. Arabinoxylo-oligosaccharides (AXOS), levan-type FOS, gentio-oligosaccharides (GenOS) and pectin-derived oligosaccharides (POS) are examples of these new potential products.

5.1. Isomalto-oligosaccharides

Isomalto-oligosaccharides are usually found as a mixture of oligosaccharides with predominantly α-(1,6)-linked glucose residues with a degree of polymerization (DP) ranging from 2–6, and oligosaccharides with a mixture of α-(1,6) and occasionally α-(1,4) glycosidic bonds such as panose (Figure 8; Table 1) [152].

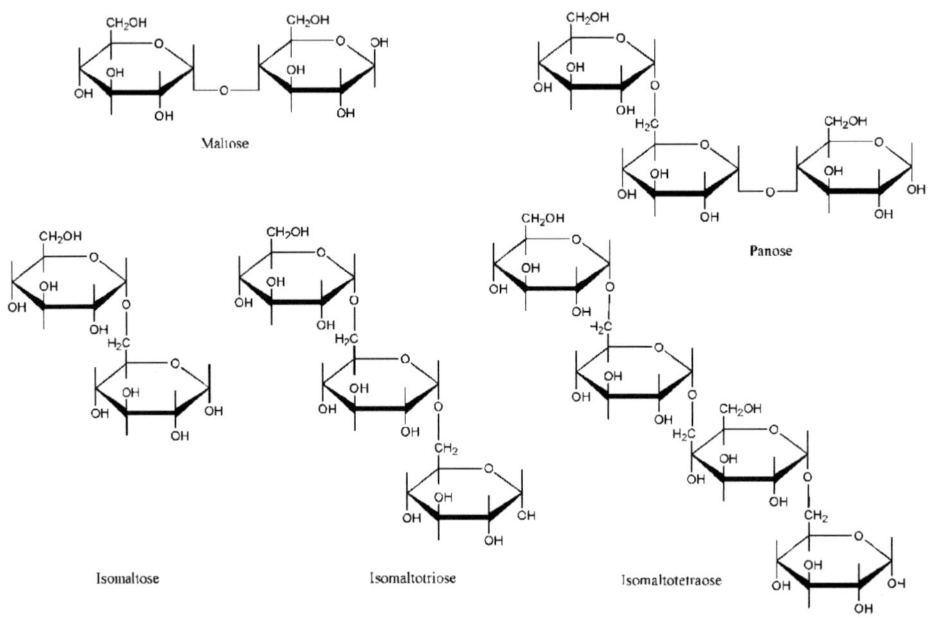

Figure 8. Examples of structures of isomalto-oligosaccharides. Glucosyl residues are linked to maltose or isomaltose by α-(1,6) glycosidic bonds.

Isomalto-oligosaccharides, like malto-oligosaccharides, are produced using starch as the raw material. Isomalto-900, a commercial product, is produced from cornstarch and consists of isomaltose, isomaltotriose and panose. Starch dextrans are easily converted to IMO, which are the market leaders in the dietary carbohydrate sector of functional foods in Japan. However, unlike malto-oligosaccharides, there is evidence to suggest that isomalto-oligosaccharides induce a bifidogenic response [11].

IMO occur naturally in various fermented foods and sugars such as sake, soybean sauce and honey. They are a product of an enzymatic transfer reaction, using a combination of immobilized enzymes. Initially, starch is liquefied using α-amylase (EC 3.2.1.1) and pullulanase (EC 3.2.1.41), and, in a second stage, the intermediary product is processed by both β-amylase (EC 3.2.1.2) and α-glucosidase (EC 3.2.1.20). Beta-amylase first hydrolyzes the liquefied starch to maltose. The transglucosidase activity of α-glucosidase then produces isomalto-oligosaccharides mixtures which contain oligosaccharides with both α-(1,6)- and α-(1,4)-linked glucose residues (Table 2) [153].

In recent years, much research has been focused on improvement of the efficiency of IMO production by screening for new and better enzymes for high yield IMO synthesis. Efforts also have been made to develop novel processes such as synthesis of IMO from sucrose using free or immobilized dextransucrase and dextranase, and efficient conversion of maltose into IMO using immobilized transglucosidase, or using an enzyme membrane reactor [153,154].

IMO are mild in taste and relatively inexpensive to produce. These oligosaccharides have desirable physicochemical characteristics such as relatively low sweetness, low viscosity and bulking properties. IMOs have been developed to prevent dental caries, as substitute sugars for diabetics [155], or to improve the intestinal flora [152].

Several companies currently manufacture isomaltooligosaccharides, of which Showa Sangyo (Japan) is the major producer. Of the emerging prebiotic oligosaccharides, IMO are used in the largest quantities for food applications. In Japan, the volume of IMOs manufactured is estimated to be three times greater than for either FOS or GOS [152]. Among other oligosaccharides, which are widely used as food ingredients or additives [156] based on their nutritional and health benefits [157], IMO are interesting due to availability, high stability and low cost [154].

Unlike other prebiotic oligosaccharides, considerable digestion of IMO occurs during intestinal transit. A large portion of this ingredient reaches the colon and intestinal enzymes degrade the remainder, leading to a rise in blood glucose levels [154]. Thus, a part of the IMO survives gastric transit to be fermented by the intestinal microbiota [152]. *In vitro* fermentation studies have shown that IMO promote the selective proliferation of bifidobacteria in the fecal microbiota [158]. However, further controlled human feeding studies employing culture and molecular techniques are required to determine the impact of IMO on the intestinal microbiota.

Beneficial effects of IMO consumption have been reported in a few human feeding studies investigating health parameters in specific populations. IMOs stimulate bowel movement and help to decrease total cholesterol levels with an intake of 10 g/d in elderly people [158].The limited data for physiological effects showed only improved defecation pattern (frequency and stool bulk via increases in microbial biomass) and lowering of total cholesterol levels [158,159]. In conclusion, the data for the bifidogenic effects of isomalto-oligosaccharides are less consistent than for other typical oligosaccharides like inulin or oligofructose [155].

5.2. Soybean oligosaccharides

Unlike other oligosaccharides, soybean oligosaccharides are extracted directly from the raw material and do not require enzymatic manufacturing processes. These α-galactooligosaccharides include and consist of galactosyl residues linked to the glucose moiety of sucrose by α-(1,6) bonds (Figure 9,Table 1) [2].

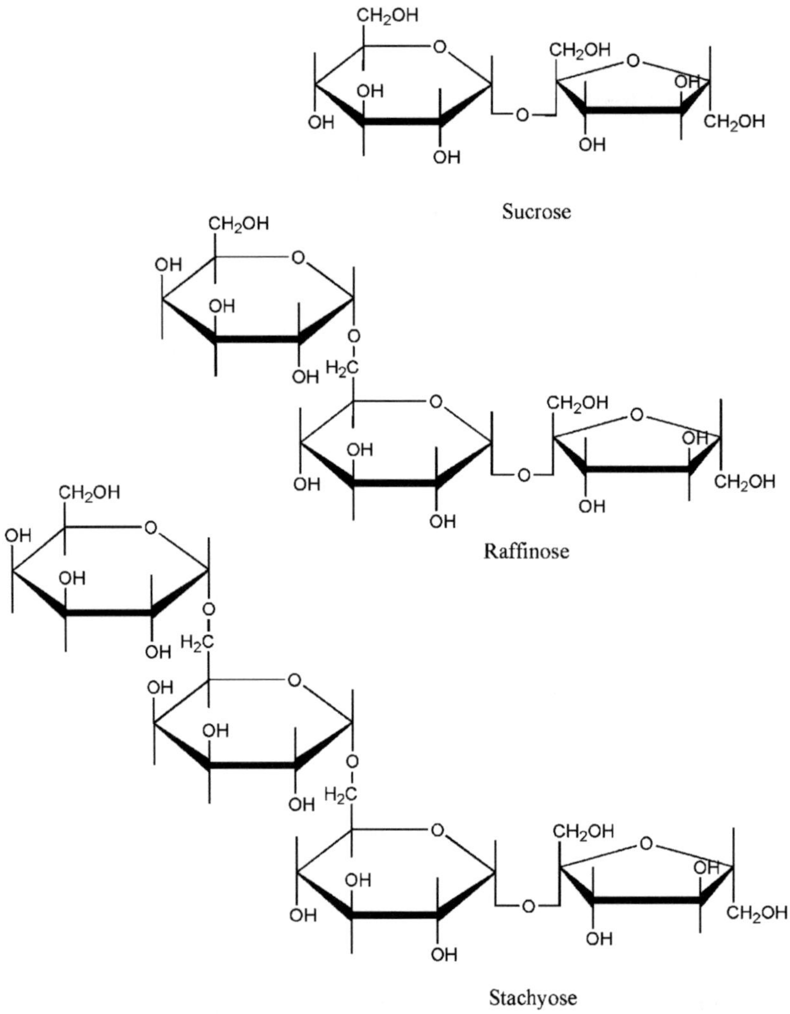

Figure 9. Examples of the main soybean oligosaccharides, raffinose and stachyose, derived from sucrose, showing galactosyl residues linked to sucrose by α-(1,6) bonds.

Soybean whey, a by-product from the production of soy protein isolates and concentrates, is composed mainly of raffinose (DP 3), stachyose (DP 4) and verbascose (DP 5), as well as sucrose, glucose and fructose. The most abundant sugars are extracted from the soybean whey and concentrated to produce soybean oligosaccharide syrup (Table 2), rather than being commercially synthesized using enzymatic processes [158].

Raffinose and stachyose are resistant to digestion, since α-galactosidase activity (required to hydrolyze these carbohydrates) is not present among human digestive enzymes and, therefore, reach the colon intact, where they act as prebiotics, stimulating the growth of bifidobacteria. Apart from being acknowledged as non-digestible, human studies on the effects of these oligosaccharides are scarce. Their physiological actions appear to be similar to the other galactooligosaccharides; they are bifidogenic and promote other effects expected from this change in colon microbiota. Calpis Co. (formerly known as Calpis Food Industry Co.) produces soybean oligosaccharides in Japan [11].

5.3. Xylo-oligosaccharides

Xylo-oligosaccharides (XOS) are sugar oligomers of xylose units linked by β-(1,4) linkages (Figure 10,Table 1). The number of xylose residues can vary from 2 to 10, but mainly consist of xylobiose, xylotriose and xylo-tetraose [152], which are found naturally in bamboo shoots, fruits, vegetables, milk and honey [160]. In addition to xylose residues, xylans are usually found in combination with arabinofuranosyl, glucopyranosyl uronic acid or its 4-O-methyl derivative (2- or 3-acetyl or phenolic substituents), resulting in branched XOS with diverse biological properties [153].

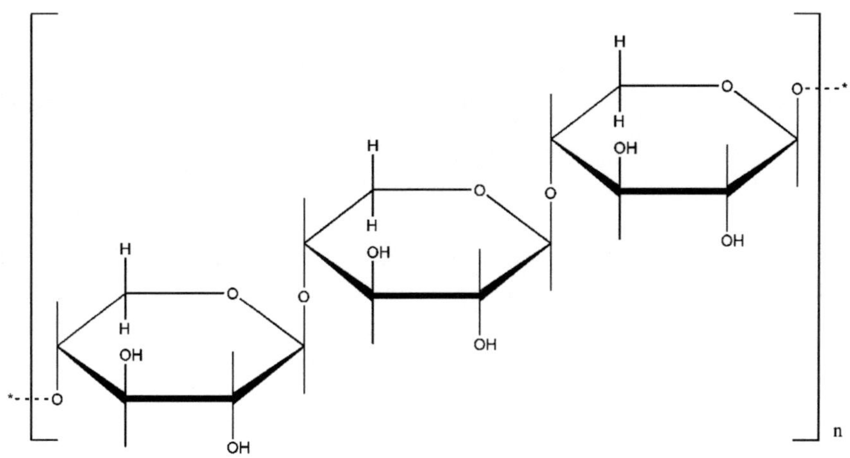

Figure 10. Partial structure of xylo-oligosaccharides (n=3-6) produced by enzymatic hydrolysis of xylan hemicelluloses, catalyzed by β-xylanases.

Their production at an industrial scale is carried out from lignocellulosic materials (LCMs). XOS can be used for several purposes, among them, food-related applications. The LCM for XOS production comes from a variety of feedstocks (from forestry, agriculture, industry or urban solid wastes) that show similarities in composition. The raw material for xylo-oligosaccharide synthesis is the polysaccharide xylan, which is extracted mainly from corncobs besides hardwoods, straws, bagasses, hulls, malt cakes and bran [160].

XOS production from LCM is not simple or economical because it depends on two treatment steps. The first step is the xylan extraction from LCM, which includes a chemical pretreatment. Although there are multiple treatments for xylan extraction (alkaline hydrolysis using NaOH, KOH, $Ca(OH)_2$, ammonia or a mixture of bases, oxidizing agents, salts or alcohols to remove lignin or pectic substances), there is no favorite consensus among them. Once the extracted xylan is in a soluble form, the second step includes the xylanase enzymatic reaction or the hydrolytic degradation of xylan by steam, water or dilute solutions of mineral acids [160]. For the enzymatic production of XOS, xylan is enzymatically hydrolysed to xylo-oligosaccharides by endo-β-1,4-xylanases (EC 3.2.1.8) (Table 2). Enzyme complexes with controlled exo-xylanase and/or β-xylosidase activity are required, to avoid the production of xylose, which may cause inhibition effects in XOS production. For food related applications, a DP of 2–4 is the most desirable [160]. Therefore, development of efficient and economical xylanase based bioprocesses for use in XOS production is necessary. Many microorganisms well known as producers of xylanolytic enzymes may be promising for novel production processes [161].

The process yields predominantly linear β-(1,4)-linked XOS (mainly xylobiose, xylotriose and xylotetraose) as well as some oligosaccharides with branched arabinose residues. For the production of food-grade XOS, a refining step is necessary. Vacuum evaporation increases the XOS concentration and removes volatile compounds such as acetic acid and the flavours of their precursors. In order to obtain higher-purity oligosaccharide products, the monosaccharide xylose and high molecular mass carbohydrates, as well as non-saccharide components can be removed from the oligosaccharides using membrane filtration techniques, organic solvent extraction, adsorption in different materials and chromatographic separation techniques used for XOS purification. Chromatographic methods, however, are not suitable for economic reasons for large-scale production of XOS intended to be used in the food industry [153].

XOS can be metabolized by bifidobacteria and lactobacilli in pure culture. In relation to human health, XOS selectively enhanced the growth of bifidobacteria thus promoting a favorable intestinal environment [152]. XOS is a promising oligosaccharide class that stimulates increased levels of bifidobacteria to a greater extent than do FOS or other oligosaccharides [161]. However, to date well-controlled animal and human feeding studies to confirm the prebiotic activity of XOS are still scarce. While they show promise, more research is required before XOS can conclusively be claimed as prebiotics. Besides the potential prebiotic effect, immunostimulating effects, antioxidant activity, anti-allergy, anti-infection and anti-inflammatory properties were reported for XOS [162-164].

In addition to the beneficial health effects, XOS have interesting physicochemical properties; they are only moderately sweet, have an acceptable odor, are noncariogenic and low caloric, stable over a wide range of pH values (2.5–8.0), even the relatively low pH value of gastric juice, and temperatures up to 100°C. Most oligosaccharides can be hydrolyzed, resulting in the loss of nutritional and physicochemical properties at acidic pH values, when treated at high temperatures for short periods, or when submitted to prolonged storage under room conditions. These properties make them suitable for incorporation into many food products such as in combination with soymilk, soft drinks, dairy products, sweets and confectionaries [158].

XOS show a remarkable potential for practical utilization in many fields, including pharmaceuticals, feed formulations and agricultural applications. Nevertheless, their most important market developments correspond to food-related applications, however, their comparatively high production costs impair market development of these oligosaccharides, and further improvements in process technology are necessary [11].

5.4. Arabinoxylo-oligosaccharides

Arabinoxylo-oligosaccharides (AXOS) are an example of a novel prebiotic dietary fiber. They can be isolated from wheat bran and consist of xylan chains with a variable substitution of arabinose side chains (Table 3) [158]. On an industrial scale, AXOS are generated through the enzymatic cleavage of AX with endoxylanases, resulting in various molecules differing in DP (between 3 and 67) and degree of substitution of arabinosyl residues [165].

Novel oligosaccharides	Chemical structure	References
Lactosucrose	4-galactosyl sucrose	[2, 156]
Arabinogalactooligosaccharides	Galactan oligomers β-(1,3) or (1,6) attached to arabinofuranose residues.	[19]
Arabinoxylooligosaccharides	Xylan randomly attached to arabinofuranose residues by α-(1,3) or α-(1,2) linkages.	[165, 237]
Arabinooligosaccharides	Arabinosyl units linked by α-(1,5) bonds.	[2]
Pectic oligosaccharides	Linear backbone of α-(1,4) linked D-galacturonic acid units randomly acetylated and/or methylated.	[171, 172, 238, 239]
Galacturonan	Linear chain of α-(1,4) linked D-galacturonic acids	
Rhamnogalacturonan	α-(1,4) linked galacturonic acid and α-(1,2) linked rhamnose units	
Mannan oligosaccharides	Mannose α-(1,6) linked backbone and α-(1,2) and α-(1,3) linked branches.	[176]
Oligodextrans	Glucosyl units linked by α-(1,4) bonds.	[176, 240]
Gentiooligosaccharides	Glucosyl units linked by β-(1,6) bonds.	
Beta-glucan oligosaccharides	Glucosyl units linked by β-(1,3/1,4) or β-(1,3/1,6) bonds.	
Cyclodextrins	α-(1,4) linked cyclic – glucosyl units.	

Table 3. Novel oligosaccharides with prebiotic activities.

The fiber properties include an improvement of bowel habit and positive change of the fermentation in the colon, whereas they were also shown to possess bifidogenic properties [158]. There are indications that AXOS have an effect against type II diabetes. AXOS decrease postprandial glucose levels and insulin response, and increase postprandial ghrelin in healthy humans [156,166].

This bifidogenic effect is strongly influenced by the complexity of the AXOS molecules and decreases with increasing average DP and degree of substitution [72,166]. Genome sequence analysis reveals that several bifidobacterial strains contain genes possibly coding for enzymes involved in the debranching of side groups and in the cleavage of the xylose backbones of AXOS [72]. This kind of specialization together with the potential to degrade xylose backbones intracellularly could explain the selective growth stimulation of bifidobacteria by AXOS.

5.5. Novel fructooligosaccharides

There is an increasing interest in novel molecules with prebiotic and physiological effects. Some fungi are able to synthesize levan-type FOS containing fructosyl units linked by β-(2,6) linkages (6-kestose being first in the series) (Table 1), or neolevan type FOS containing a fructosyl unit also linked by this type of linkage to a glucose (neokestose, neonystose, or neofructofuranosylnystose). Such FOS have been metabolized by different bifidobacteria strains when supplied as the sole carbon source [167].

Levan-type FOS were synthesized by acid hydrolysis of β-(2,6)-linked polymers containing a glucose at one terminus (levans), these have been produced by several microorganisms growing in sucrose-based medium [168]. The discovery of novel enzymes that synthesize β-(2,6)-linked FOS from sucrose may, however, provide a non-pollutant alternative to acid hydrolysis of levans. Because there is an existing process to produce inulin-type FOS, an enzymatic method involving the hydrolysis of levan to produce levan-FOS may be possible. However, with the lack of an available plant source of levan, as there is for inulin, it is possible to derive an enzymatic process to produce levan-type FOS from microbial levan, using levansucrase (Table 2) and endolevanases [169].

Marx *et al.* 2000 [170] observed that levan-type FOS obtained via the acid hydrolysis of levans were metabolized by different bifidobacteria strains, thus further demonstrating their prebiotic potential. Nevertheless, the levan-type FOS prebiotic properties have not been fully characterized, possibly due to their limited availability.

The production of levan-type FOS has not reached industrial levels [171], despite several reports demonstrating their potential applications as food and feed additives in agriculture as well as their pharmaceutical applications.

5.6. Pectic oligosaccharides

Pectic oligosaccharides (POS) (Table 3) are obtained by pectin depolymerization. Pectins are ramified heteropolymers made up of a linear backbone of α-(1,4)-linked D-galacturonic acid units (which can be randomly acetylated and/or methylated).

POS have been proposed as a new class of prebiotics capable of exerting a number of health-promoting effects. Among these are protection of colonic cells against pathogenic microorganisms [172], stimulation of apoptosis of human colonic adenocarcinoma cells [173] and *in vivo* synergistic empowerment of immunomodulation caused by galactooligosaccharides (GalOS) and fructooligosaccharides (FOS). Other benefits include potential for cardiovascular protection *in vivo*, reduction of damage by heavy metals, antiobesity effects, dermatological applications and antitoxic, antiinfection, antibacterial and antioxidant properties. Additionally, *in vivo* and *in vitro* studies have confirmed that acidic POS are not cytotoxic or mutagenic, being suitable for use in foods for children and babies [173].

5.7. Gentio-oligosaccharides

Gentio-oligosaccharides (GenOS) consist of 2–5 glucose residues linked by β-(1,6) glycosidic linkages (Table 3). These oligosaccharides are not hydrolysed in the stomach or small intestine and therefore reach the colon intact, thus fulfilling a criterion of a prebiotic [11]. GenOS were further reported to possess bifidogenic activity [153]. GenOS are usually produced from glucose syrup by enzymatic transglucosylation or by biocatalytic glycosylation with cultured cells. Despite the prebiotic potential of GenOS, research on the novel production of GenOS is sparse. Gentio-oligosaccharides are produced in Japan by Nihon Shokuhin Kako [11].

6. Perspectives

Function and application of chitooligosacharides frequently depend on their size, and, therefore, the degrees of polymerization and acetylation. Substrate-enzyme synergisms determine the molecular weight of the generated COS. Gutierrez-Román *et al.* 2014 [174] tested three chitolytic enzymes ChiA, ChiB and ChiC, alone and in combination. In addition, three chitanases were tested in synergism with a chitobiase and a non-catalytic binding protein. When evaluated individually, ChiA was unable to hydrolyze chitin while ChiB and ChiC were able to degrade chitin and generate chitin monomers and dimers. When enzymes were tested pairwise (ChiA-ChiB, ChiA-ChiC, and ChiB-ChiC) the production of dimers and trimers was much higher, and monomers significantly lower than those seen with ChiB or ChiC individually. However, higher concentrations of COS were obtained when the authors tested the four enzymes in combination with non-catalytic binding protein acting on chitin.

Further studies must be focused on the action of the enzymes on substrates with different degrees of polymerization and acetylation and N-acetylation pattern to improve the comprehension of that synergism. In addition, researches involving synergism of non-catalytic binding proteins and hydrolytic enzymes should be developed in order to increase the understanding of oligomers syntheses [127]. Consequently, to produce size-specific chitooligosaccharides by enzymatic hydrolysis, further studies on genetic modification are necessary to overproduce enzymes and non-catalytic binding proteins, which will have a great impact on the quality of oligomers obtained and on the productivity of industrial processes.

Another important challenge in the development of biotechnological processes that employ agro-food industry residues as raw material is the direct fermentation of those raw materials. Obviously, direct fermentation of raw materials is closely related with the aforementioned aspects, since fermentative processes involve microbial growth and enzymatic hydrolysis, and process conditions that in many cases are different from physiological conditions. Moreover, it is important to give attention to screening of new enzymes from extremophile microorganisms, which usually catalyze reactions under non-physiological conditions such as high salinity, high temperature and low water activity [175].

As important part of the biotechnological process, bioreactors and enzyme (free or immobilized) are essential and need special attention to improve yields and productivities. Free enzymes in batch systems are the most conventional technology employed in the production of oligosaccharides by enzymatic hydrolysis. However, it has several important drawbacks, because enzymes are unstable, can be employed once and accumulation of products usually reduces their activity. These drawbacks are related directly to the quality of the product and the yield of the process. Development of novel technologies in order to solve those snags employing immobilized enzymes in column reactor and membrane systems have been studied. Column reactor packing with immobilized enzymes allows continuous production of oligosaccharides and has important advantages, such as increased operational stability of the enzyme and reduced accumulation which otherwise could lead to enzyme inhibition. The poorer affinity of immobilized enzymes is the main disadvantage of the application of column reactors at industrial scale. Studies should be directed towards the improvement of enzyme-support affinity. Membrane reactors are considered a new and attractive technology to produce oligosaccharides, in which enzymes are confined in the reaction side and continuously reused, with obvious implications for the efficiency and economy of the process. Low-cost and low-energy consumption are other important advantages to increase its utilization. The main limitation for industrial application of membrane reactors are fouling and polarization phenomena, which decrease considerably permeate flux, containing the produced oligosaccharides [176]. The main challenge to be studied in order to implement this technology advantageously in the industry is how to reduce the effect of these problems without affecting the stability of enzymes.

Author details

Tathiana Souza Martins Meyer, Ângelo Samir Melim Miguel,
Daniel Ernesto Rodríguez Fernández and Gisela Maria Dellamora Ortiz*

*Address all correspondence to: gisela@pharma.ufrj.br

School of Pharmacy, Federal University of Rio de Janeiro, Rio de Janeiro, Brazil

References

[1] Miguel, A S M, Martins-Meyer, T S, Figueiredo, E V C, Lobo, B W P, Dellamora-Ortiz, G M. Enzymes in Bakery: Current and Future Trends. In: Muzzalupo, I, editor. Food Industry. 1ed. Rijeka: InTech; 2013. p. 287-321. DOI: 10.5772/53168.ch14

[2] Patel S, Goyal A. Functional oligosaccharides: production, properties and applications. World Journal of Microbiology and Biotechnology. 2011;27:1119-1128. DOI: 10.1007/s11274-010-0558-5

[3] Villares, J M M. Prebiotics in Infant Formulas: Risks and Benefits. In: Watson, R R, Preedy, V R, editors. Bioactive Foods in Promoting Health. London: Academic Press; 2010. p. 117-129. ch8

[4] Pinelo M, Jonsson G, Meyer AS. Membrane technology for purification of enzymatically produced oligosaccharides: Molecular and operational features affecting performance. Separation and Purification Technology. 2009;70:1-11. DOI:10.1016/j.seppur.2009.08.010

[5] Gibson, G R, Probert, H M, Loo, J V, Rastall, R A, Roberfroid, M B. Dietary modulation of the human colonic microbiota: updating the concept of prebiotics. Nutrition Research Reviews. 2004;17:259-275. DOI: 10.1079/NRR200479

[6] Roberfroid, M B. Prebiotics: Concept, Definition, Criteria, Methodologies, and Products. In: Gibson, G R, Roberfroid, M B, editors. Handbook of Prebiotics. Boca Raton: CRC Press; 2008. p. 39-68. Ch. 4.

[7] Rioux, K P, Madsen, K L, Fedorak, R N. The Role of Enteric Microflora in Inflammatory Bowel Disease: Human and Animal Studies with Probiotics and Prebiotics. Gastroenterology Clinics of North America. 2005;34:465-482. DOI: 10.1016/j.gtc.2005.05.005

[8] Saad N, Delattre C, Urdaci M, Schmitter J M, Bressollier P. An overview of the last advances in probiotic and prebiotic field. LWT - Food Science and Technology. 2013;50:1-16. DOI: 10.1016/j.lwt.2012.05.014

[9] Scheid, M M A, Moreno, Y M F, Marostica-Junior, M R, Pastore, G M. Effect of prebiotics on the health of the elderly. Food Research International. 2013;53:426-432. DOI: 10.1016/j.foodres.2013.04.003

[10] Bruins, M E. Oligosaccharide Producion with Thermophilic Enzymes [thesis]. Wageningen: Wageningen University; 2003.

[11] Crittenden, R G, Playne, M J. Production, properties and applications of food-grade oligosaccharides. Trends in Food Science and Technology. 1996;7:353-361. DOI: 10.1016/S0924-2244(96)10038-8

[12] Mussatto, S I, Mancilha, I M. Non-digestible oligosaccharides: A review. Carbohydrate Polymers. 2007;68:587-597. DOI: 10.1016/j.carbpol.2006.12.011

[13] Courtois, J. Oligosaccharides from land plants and algae: production and applications in therapeutics and biotechnology. Current Opinion in Microbiology. 2009;12:261-273. DOI 10.1016/j.mib.2009.04.007

[14] Grout, D H G, Gabin, V. Glycosidases and glycosyl transferases in glycoside and oligosaccharide synthesis. Current Opinion in Chemical Biology. 1998;2:98-111. DOI: 10.1016/S1367-5931(98)80041-0

[15] Casci, T, Rastall, R A. Manufacture of Prebiotic Oligosaccharides. In: Gibson, G R, Rastall, R A, editors. Prebiotics: Development and Application. Chichester: John Wiley & Sons, Ltd.; 2006. p. 29-55. Ch 2.

[16] Patel, V, Saunders, G, Bucke, C. Production of fructooligosaccharides by *Fusarium oxysporum*. Biotechnology Letters. 1994;16:1139-1144. DOI: 10.1007/BF01020840

[17] Viikari, L, Gisler, R. By-products in the fermentation of sucrose by different *Zymomonas*-strains. Applied Microbiology and Biotechnology. 1986;23:240-244. DOI: 10.1007/BF00261922

[18] Van Loo, J, Coussement, P, De Leenheer, L, Hoebregs, H, Smits, G. On the presence of inulin and oligofructose as natural ingredients in the Western diet. Critical Reviews in Food Science and Nutrition. 1995;35:525-552. DOI: 10.1080/10408399509527714

[19] Barreteau, H, Delattre, C, Michaud, P. Production of oligosaccharides as promising new food additive generation. Food Technology and Biotechnology. 2006;44:323-333.

[20] Madlova, A, Antosova, M, Barathova, M, Polakovic, M, Stefuca, V, Bales, V. Biotransformation of sucrose to fructooligosaccharides: the choice of microorganisms and optimization of process conditions. Progress in Biotechnology. 2000;17:151-155. DOI: 10.1016/S0921-0423(00)80061-1

[21] Silva, M F, Rigo, D, Mossi, V, Golunski, S, Kuhn, G O, Di Luccio, M, Dallago, R, Oliveira, D, Oliveira, J V, Treichel, H. Enzymatic synthesis of fructooligosaccharides by inulinases from *Aspergillus niger* and *Kluyveromyces marxianus* NRRL Y-7571 in aqueous–organic medium. Food Chemistry. 2013;138:148-153. DOI: 10.1016/j.foodchem.2012.09.118

[22] Roberfroid, M B. Functional foods: Concepts and application to inulin and oligofructose. British Journal of Nutrition. 2002;87:S139-S143. DOI: 10.1079/BJN/2002529

[23] L'homme, C, Puigserver, A, Biagini, A. Effect of food-processing on the degradation of fructooligosaccharides in fruit. Food Chemistry. 2003;82:533-537. DOI: 10.1016/S0308-8146(03)00003-7

[24] Sabater-Molina, M, Larqué, E, Torrella, F, Zamora, S. Dietary fructooligosaccharides and potential benefits on health. Journal of Physiology and Biochemistry. 2009;65:315-328. DOI: 10.1007/BF03180584

[25] Vega, R, Zuniga-Hansen, M E. The effect of processing conditions on the stability of fructooligosaccharides in acidic food products. Food Chemistry. 2015;173:784-789. DOI: 10.1016/j.foodchem.2014.10.119

[26] Birkett, A M, Francis, C C. Short-Chain Fructo-Oligosaccharide. A Low Molecular Weight Fructan. In: Cho, S S, Finocchiaro, E T, editors. Handbook of Prebiotics and Probiotics Ingredients: Health Benefits and Food Applications. Boca Raton: CRC Press; 2010. p. 13-42. Ch. 2.

[27] Yun, J W. Fructooligosaccharides – Occurrence, preparation and application. Enzyme and Microbial Technology. 1996;19:107-117. DOI: 10.1016/0141-0229(95)00188-3

[28] Ziemer, C J, Gibson, G R. An overview of probiotics, prebiotics and synbiotics in the functional food concept: perspectives and future strategies. International Dairy Journal. 1998;8:473-479. DOI: 10.1016/S0958-6946(98)00071-5

[29] Sangeetha, P T, Ramesh, M N, Prapulla, S G. Recent trends in the microbial production, analysis, and application of fructooligosaccharides. Trends in Food Science and Technology. 2005;16:442-457. DOI: 10.1016/j.tifs.2005.05.003

[30] Sangeetha, P T, Ramesh, M N, Prapulla, S G. Maximization of fructooligosaccharide production by two stage continuous process and its scale up. Journal of Food Engineering. 2005;68:57-64. DOI: 10.1016/j.jfoodeng.2004.05.022

[31] Roberfroid, M B. Inulin-type fructans: Functional food ingredients. Journal of Nutrition. 2007;137:2493S-2502S.

[32] Silva, M F, Rigo, D, Mossi, V, Golunski, S, Kuhn, G O, Di Luccio, M, Dallago, R, Oliveira, D, Oliveira, J V, Treichel, H. Enzymatic synthesis of fructooligosaccharides by inulinases from *Aspergillus niger* and *Kluyveromyces marxianus* NRRL Y-7571 in aqueous–organic medium. Food Chemistry. 2013;138:148-153. DOI: 10.1016/j.foodchem.2012.09.118

[33] Arrizon, J, Urias-Silvas, J E, Sandoval, G, Mancilla-Margalli, N A, Gschaedler, A C, Morel, S, Monsan, P. Production and Bioactivity of Fructan-Type Oligosaccharides. In: Moreno, F J, Sanz, M L, editors. Food Oligosaccharides: Production, Analysis and Bioactivity. Chichester: John Wiley & Sons, Ltd.; 2014. p. 184-199. Ch. 11.

[34] Chen, W-C, Liu, C-H. Production of β-fructofuranosidase by *Aspergillus japonicus*. Enzyme and Microbial Technology. 1996;18:153-160. DOI: 10.1016/0141-0229(95)00099-2

[35] Ganaie, M A, Gupta, U S, Kango, N. Screening of biocatalysts for transformation of sucrose to fructooligosaccharides. Journal of Molecular Catalysis B: Enzymatic. 2013;97:12-17. DOI: 10.1016/j.molcatb.2013.07.008

[36] Antosova, M, Polakovic, M. Fructosyltransferases: The enzymes catalyzing production of fructooligosaccharides. Chemical Papers. 2001;55:350-358.

[37] Kim M-H, In M-J, Cha H J, Yoo Y J. An empirical rate equation for the fructooligosaccharide-producing reaction catalyzed by β-fructofuranosidase. Journal of Fermentation and Bioengineering. 1996;82:458-463. DOI: 10.1016/S0922-338X(97)86983-8

[38] Fernandez, R C, Ottoni, C A, Silva, E S, Matsubara, R M S, Carter, J M, Magossi, L R, Wada, M A A, Rodrigues, M F A, Maresma, B G, Maiorano, A E. Screening of β-fructofuranosidase-producing microorganisms and effect of pH and temperature on enzymatic rate. Applied Microbiology and Biotechnology. 2007;75:87-93. DOI: 10.1007/s00253-006-0803-x

[39] Perez, E R, Trujillo, L E, Arrieta, J G, Pérez, H, Brizuela, M A, Trujillo, G, Hernández, L. A pH shift-based procedure to screen fructooligosaccharides fermenting yeast or bacterial strains. Biotecnología Aplicada. 2010;27:216-220.

[40] Maugeri, F, Hernalsteens, S. Screening of yeast strains for transfructosylating activity. Journal of Molecular Catalysis B: Enzymatic. 2007;49:43-49. DOI: 10.1016/j.molcatb.2007.08.001

[41] Zambelli, P, Fernandez-Arrojo, L, Romano, D, Santos-Moriano, P, Gimeno-Perez, M, Poveda, A, Gandolfi, R, Fernández-Lobato, M, Molinari, F, Plou, F J. Production of fructooligosaccharides by mycelium-bound transfructosylation activity present in *Cladosporium cladosporioides* and *Penicillium sizovae.* Process Biochemistry. 2014;49:2174-2180. DOI: 10.1016/j.procbio.2014.09.021

[42] Antosova, M, Illeova, V, Vandakova, M, Druzkovska, A, Polakovic, M. Chromatographic separation and kinetic properties of fructosyltransferase from *Aureobasidium pullulans.* Journal of Biotechnology. 2008;135:58-63. DOI: 10.1016/j.jbiotec.2008.02.016

[43] Maiorano, A E, Piccoli, R M, Silva, E S, Rodrigues, M F A. Microbial production of fructosyltransferases for synthesis of pre-biotics. Biotechnol Letters. 2008;30:1867-1877. DOI: 10.1007/s10529-008-9793-3

[44] Nemukula, A, Mutanda, T, Wilhelmi, B S, Whiteley, C G. Response surface methodology: Synthesis of short chain fructooligosaccharides with a fructosyltransferase from *Aspergillus aculeatus*. Bioresource Technology. 2009;100:2040-2045. DOI: 10.1016/j.biortech.2008.10.022

[45] Yoshikawa, J, Amachi, S, Shinoyama, H, Fujii, T. Purification and some properties of β-fructofuranosidase I formed by *Aureobasidium pullulans* DSM 2404. Journal of Bioscience and Bioengineering. 2007;103:491-493. DOI: 10.1263/jbb.103.491

[46] Yoshikawa, J, Amachi, S, Shinoyama, H, Fujii, T. Production of fructooligosaccharides by crude enzyme preparations of β-fructofuranosidase from *Aureobasidium pullulans*. Biotechnology Letters. 2008;30:535-539. DOI: 10.1007/s10529-007-9568-2

[47] Jung, K H, Yun, J W, Kang, K R, Lim, J Y, Lee, J H. Mathematical model for enzymatic production of fructo-oligosaccharides from sucrose. Enzyme and Microbial Technology. 1989;11:491-494. DOI: 10.1016/0141-0229(89)90029-X

[48] Duan, K J, Chen, J S, Sheu, D C. Kinetic studies and mathematical model for enzymatic production of fructooligosaccharides from sucrose. Enzyme and Microbial Technology. 1994;16:334-339. DOI: 10.1016/0141-0229(94)90176-7

[49] Sheu, D C, Lio, P J, Chen, S T, Lin, C T., Duan, K J. Production of fructooligosaccharides in high yield using a mixed enzyme system of β-fructofuranosidase and glucose oxidase. Biotechnology Letters. 2001;23:1499-1503. DOI: 10.1023/A:1011689531625

[50] Vega, R, Zuniga-Hansen, M E. Enzymatic synthesis of fructooligosaccharides with high 1-kestose concentrations using response surface methodology. Bioresource Technology. 2011;102:10180-10186. DOI:10.1016/j.biortech.2011.09.025

[51] Hang, Y D, Woodams, E E. Fructosyltransferase activity of commercial enzyme preparations used in fruit juice processing. Biotechnology Letters. 1995;17:741-745. DOI: 10.1007/BF00130361

[52] Tanriseven, A, Gokmen, F. Novel method for the production of a mixture containing fructooligosaccharides and isomaltooligosaccharides. Biotechnology Techniques. 1999;13:207-210. DOI: 10.1023/A:1008961016065

[53] Ghazi, I, Fernandez-Arrojo, L, Gomez De Segura, A, Alcalde, M, Plou, F J, Ballesteros, A. Beet sugar syrup and molasses as low-cost feedstock for the enzymatic production of fructo-oligosaccharides. Journal of Agricultural and Food Chemistry. 2006;54:2964-2968. DOI: 10.1021/jf053023b

[54] Madlova, A, Antosova, M, Barathova, M, Polakovic, M, Stefuca, V, Bales, V. Screening of microorganisms for transfructosylating activity and optimization of biotransformation of sucrose to fructooligosacharides. Chemical Papers. 1999;53:366-369.

[55] Vega-Paulino, R J, Zuniga-Hansen, M E. Potential application of commercial enzyme preparations for industrial production of short-chain fructooligosaccharides. Journal of Molecular Catalysis B: Enzymatic. 2012;76:44-51. DOI: 10.1016/j.molcatb.2011.12.007

[56] Guisan, J M. Immobilization of Enzymes as the 21st Century Begins. An Already Solved Problem or Still an Exciting Challenge? In: Guisan, J M, editor. Immobilization of Enzymes and Cells. Second Edition. Totowa: Humana Press; 2006. p. 1-13. ch1.

[57] Sheldon, R A, van Pelt, S. Enzyme immobilisation in biocatalysis: why, what and how. Chemical Society Reviews. 2013;42:6223-6235. DOI: 10.1039/c3cs60075k

[58] Tanriseven, A, Aslan, Y. Immobilization of Pectinex Ultra SP-L to produce fructooligosaccharides. Enzyme and Microbial Technology. 2005;36:550-554. DOI: 10.1016/j.enzmictec.2004.12.001

[59] Ghazi, I, De Segura, A G, Fernandez-Arrojo, L, Alcalde, M, Yates, M, Rojas-Cervantes, M L, Plou, F J, Ballesteros, A. Immobilisation of fructosyltransferase from *Aspergillus aculeatus* on epoxy-activated Sepabeads EC for the synthesis of fructo-oligosaccharides. Journal of Molecular Catalysis B: Enzymatic. 2005;35:19-27. DOI: 10.1016/j.molcatb.2005.04.013

[60] Fernandez-Arrojo, L, Rodriguez-Colinas, B, Gutierrez-Alonso, P, Fernandez-Lobato, M, Alcalde, M, Ballesteros, A O, Plou, F J. Dried alginate-entrapped enzymes (DALGEEs) and their application to the production of fructooligosaccharides. Process Biochemistry. 2013;48:677-682. DOI: 10.1016/j.procbio.2013.02.015

[61] Lorenzoni, A S G, Aydos, L F, Klein, M P, Rodrigues, R C, Hertz, P F. Fructooligosaccharides synthesis by highly stable immobilized β-fructofuranosidase from *Aspergillus aculeatus*. Carbohydrate Polymers. 2014;103:193-197. DOI: 10.1016/j.carbpol.2013.12.038

[62] Dominguez, A, Nobre, C, Rodrigues, L R, Peres, A M, Torres, D, Rocha, I, Lima, N, Teixeira, J. New improved method for fructooligosaccharides production by *Aureobasidium pullulans*. Carbohydrate Polymers. 2012;89:1174-1179. DOI: 10.1016/j.carbpol.2012.03.091

[63] Chien, C-S, Lee, W-C, Lin, T-J. Immobilization of *Aspergillus japonicus* by entrapping cells in gluten for production of fructooligosaccharides. Enzyme and Microbial Technology. 2001;29:252-257. DOI: 10.1016/S0141-0229(01)00384-2

[64] Mussatto, S I, Aguilar, C N, Rodrigues, L R, Teixeira, J A. Fructooligosaccharides and β-fructofuranosidase production by *Aspergillus japonicus* immobilized on lignocellulosic materials. Journal of Molecular Catalysis B: Enzymatic. 2009;59:76-81. DOI: 10.1016/j.molcatb.2009.01.005

[65] Cheng, C Y, Duan, K J, Sheu, D C, Lin, C T, Li, S Y. Production of fructooligosaccharides by immobilized mycelium of *Aspergillus japonicus*. Journal of Chemical Technology and Biotechnology. 1996;66:135-138. DOI: 10.1002/(SICI)1097-4660(199606)66:2<135::AID-JCTB479>3.0.CO;2-S

[66] Cruz, R, Cruz, V D, Belini, M Z, Belote, J G, Vieira, C R. Production of fructooligosaccharides by the mycelia of *Aspergillus japonicus* immobilized in calcium alginate. Bioresource Technology. 1998;65:139-143. DOI: 10.1016/S0960-8524(98)00005-4

[67] Charalampopoulos D, Rastall R A. Prebiotics in foods. Current Opinion in Biotechnology. 2012;23:187-191. DOI; 10.1016/j.copbio.2011.12.028

[68] Lopez, H W, Coudray, C, Levrat-Verny, M-A, Feillet-Coudray, C, Demigne, C, Remesy, C. Fructooligosaccharides enhance mineral apparent absorption and counteract the deleterious effects of phytic acid on mineral homeostasis in rats. Journal of Nutrition and Biochemistry. 2000;11:500-508. DOI: 10.1016/S0955-2863(00)00109-1

[69] Biedrzycka, E, Bielecka, M. Prebiotic effectiveness of fructans of different degrees of polymerization. Trends in Food Science and Technology. 2004;15:170-175. DOI: 10.1016/j.tifs.2003.09.014

[70] Kaplan, H, Hutkins, R W. Fermentation of fructooligosaccharides by lactic acid bacteria and bifidobacteria. Applied and Environmental Microbiology. 2000;66:2682-2684.

[71] Kaplan, H, Hutkins, R W. Metabolism of fructooligosaccharides by *Lactobacillus paracasei* 1195. Applied and Environmental Microbiology. 2003;69:2217-2222. DOI: 10.1128/AEM.69.4.2217-2222.2003

[72] De Vuyst, L, Moens, F, Selak, M, Riviere, A, Leroy, F. Summer Meeting 2013: growth and physiology of bifidobacteria. Journal of Applied Microbiology. 2013;116:477-491. DOI: 10.1111/jam.12415

[73] Mabel, M J, Sangeetha, P T, Platel, K, Srinivasan, K, Prapulla, S G. Physicochemical characterization of fructooligosaccharides and evaluation of their suitability as a potential sweetener for diabetics. Carbohydrate Research. 2008;343:56-66. DOI: 10.1016/j.carres.2007.10.012

[74] Marx, S P, Winkler, S, Hartmeier, W. Metabolization of beta-(2,6)-linked fructose-oligosaccharides by different bifidobacteria. FEMS Microbiology Letters 2000;182:163-169. DOI: 10.1111/j.1574-6968.2000.tb08891.x

[75] Grizard, D, Barthomeuf, C. Enzymatic synthesis and structure determination of NEO-FOS. Food Biotechnology. 1999;13:93-105. DOI: 10.1080/08905439609549963

[76] Holsinger, V H. Lactose. In: Wong, N P, Jenness, R, Keeney, M, Marth, E H, editors. Fundamentals of Dairy Chemistry. 3rd ed. New York: Van Nostrand Reinhold Co.; 1988. p. 279-342.

[77] Yang, S T, Silva, E M. Novel products and new technologies for use of a familiar carbohydrate, milk lactose. Journal of Dairy Science. 1995;78:2541-2562.

[78] Paterson, A H J. Production and uses of lactose. In: McSweeney, P L H, Fox, P F, editors. Lactose, Water, Salts and Minor Constituents. 3rd ed. New York: Springer; 2009. p 105-120.

[79] Fox, P F. Lactose: Chemistry and Properties. In: McSweeney, P L H, Fox, P F, editors. Lactose, Water, Salts and Minor Constituents. 3rd ed. New York: Springer; 2009. p.1–15. DOI: 10.1007/978-0-387-84865-5Fox 2009

[80] Schaafsma, G. Lactose and lactose derivatives as bioactive ingredients in human nutrition. International Dairy Journal. 2008;18:458-465. DOI: 10.1016/j.idairyj.2007.11.013Schaafsma 2008

[81] Paige, D M. Lactose Intolerance. In: Caballero, B, Allen, L, Prentice, A, editors. Encyclopedia of Human Nutrition. 2nd ed. Oxford: Elsevier Ltd.; 2005. p. 113-120.

[82] Mahoney, R R. Galatosyl-oligosaccharide formation during lactose hydrolysis: a review. Food Chemistry. 1998;63:147-154. DOI: 10.1016/S0308-8146(98)00020-X

[83] Wallenfels, K, Malhotra, O P. Beta-galactosidase. In: Boyer, P D, editor. The Enzymes. 2nd ed. New York: Academic Press Inc.; 1960. p. 409-430.

[84] Lomer, M C E, Parkes, G C, Sanderson, J D. Review article: Lactose intolerance in clinical practice – Myths and realities. Alimentary Pharmacology and Therapeutics. 2008;27:93-103. DOI: 10.1111/j.1365-2036.2007.03557.x

[85] Tzortzis, G, Vulevic, J. Galacto-oligosaccharide Prebiotics. In: Charalampopoulos, D, Rastall, R A, editors. Prebiotics and Probiotics Science and Technology. New York: Springer; 2009. p. 207-244.

[86] Villamiel, M, Montilla, A, Olano, A, Corzo, N. Production and Bioactivity of Oligosaccharides Derived from Lactose. In: Moreno, F J, Sanz, M L, editors. Food Oligosaccharides: Production, Analysis and Bioactivity. John Wiley & Sons. 2014, p. 137.

[87] Oliveira, C, Guimarães, P M R, Domingues, L. Recombinant microbial systems for improved β-galactosidase production and biotechnological applications. Biotechnology Advances. 2011;29:600-609. DOI: 10.1016/j.biotechadv.2011.03.008

[88] Torres, D P M, Gonçalves, M F, Teixeira, J A, Rodrigues, L R. Galactooligosaccharides: production, properties, applications, and significance as prebiotics. Comprehensive Reviews in Food Science and Food Safety. 2010;9:438-454. DOI: 10.1111/j. 1541-4337.2010.00119.x

[89] Martinez-Villaluenga, C, Cardelle-Cobas, A, Corzo, N, Olano, A, Villamiel, M. Optimization of conditions for galactooligosaccharides synthesis during lactose hydrolysis by β-galactosidase from *Kluyveromyces lactis* (Lactozym 3000 L HP G). Food Chemistry. 2008;107:258-264. DOI: 10.1016/j.foodchem.2007.08011

[90] Hatzinikolaou, D G, Katsifas, E, Mamma, D, Karagouni, A D, Christakopoulos, P, Kekos, D. Modeling of the simultaneous hydrolysis–ultrafiltration of whey permeate by a thermostable β-galactosidase from *Aspergillus niger*. Biochemical Engineering Journal. 2005;24:161-172. DOI: 10.1016/j.bej.2005.02.011

[91] Jurado, E, Camacho, F, Luzón, G, Vicaria, J M. Kinetic models of activity for β-galactosidases: influence of pH, ionic concentration and temperature. Enzyme and Microbial Technology. 2004;34:33-40. DOI: 10.1016/j.enzmictec.2003.07.004

[92] Buchholz, K, Kasche, V, Bornscheuer, U T. Equilibrium and Kinetically Controlled Reactions Catalysed by Enzymes. In: Biocatalysts and Enzyme Technology. Weinheim: Wiley-VCH Verlag GmbH& Co.; 2005. p. 42-46.

[93] Chen, S-X, Wei, D-Z, Hu, Z-H. Synthesis of galacto-oligosaccharides in AOT/isooctane reverse micelles by beta-galactosidase. Journal of Molecular Catalysis B: Enzymatic. 2001;16:109-114. DOI: 10.1016/S1381-1177(01)00051-0

[94] Chen, C W, Ou-Yang, C C, Yeh, C W. Synthesis of galactooligosaccharides and transgalactosylation modeling in reverse micelles. Enzyme and Microbial Technology. 2003;33:497-507. DOI: 10.1016/S0141-0229(03)00155-8

[95] Roos, Y H. Solid and Liquid States of Lactose. In: McSweeney, P L H, Fox, P F, editors. Lactose, Water, Salts and Minor Constituents. 3rd ed. New York: Springer; 2009. p. 17-33.

[96] Bruins, M E, van Hellemond, E W, Janssen, A E M, Boom, R M. Maillard reactions and increased enzyme inactivation during oligosaccharide synthesis by a hyperther-

mophilic glycosidase. Biotechnology and Bioengineering. 2003;81:546-552. DOI: 10.1002/bit.10498

[97] Bruins, M E, Janssen, A E M, Boom, R M. Thermozymes and their applications—a review of recent literature and patents. Applied Biochemistry and Biotechnology. 2001;90:155-186. DOI: 10.1385/ABAB:90:2:155

[98] Boon, M A, van der Oost, J, de Vos, A E M, van't Riet, K. Synthesis of oligosaccharides catalyzed by thermostable β-glucosidase from *Pyrococcus furiosus*. Applied Biochemistry and Biotechnology. 1998;75: 269-278. DOI: 10.1007/BF02787780

[99] Chen, W, Chen, H, Xia, Y, Yang, J, Zhao, J, Tian, F, Zhang, H P, Zhang, H. Immobilization of recombinant thermostable β-galactosidase from *Bacillus stearothermophilus* for lactose hydrolysis in milk Journal of Dairy Science. 2009;92:491-498. DOI: 10.3168/jds.2008-1618

[100] Petzelbauer, I, Reiter, A, Splechtna, B, Kosma, P, Nidetzky, B. Transgalactosylation by thermostable β-glycosidases from *Pyrococcus furiosus* and *Sulfolobus solfataricus*. Binding interactions of nucleophiles with the galactosylated enzyme intermediate makes major contributions to the formation of new beta-glycosides during lactose conversion. European Journal of Biochemistry. 2000;267:5055-5066. DOI: 10.1046/j.1432-1327.2000.01562.x

[101] Petzelbauer I, Zeleny R, Reiter A, Kulbe D, Nidetzky B. Development of an ultra-high-temperature process for the enzymatic hydrolysis of lactose: II. Oligosaccharide formation by two thermostable β-glycosidases. Biotechnology and Bioengineering. 2000;69:140-149. DOI: 10.1002/(SICI)1097-0290(20000720)692<140::AID-BIT3>3.0.CO;2-R

[102] Reuter, S, Rusborg, N A, Zimmermann, W. β-Galactooligosaccharide synthesis with β-galactosidases from *Sulfolobus solfataricus*, *Aspergillus oryzae*, and *Escherichia coli*. Enzyme and Microbial Technology. 1999;25:509-516. DOI: 10.1016/S0141-0229(99)00074-5

[103] Chen, W, Chen, H, Xia, Y, Zhao, J, Tian, F, Zhang, H. Production, purification, and characterization of a potential thermostable galactosidase for milk lactose hydrolysis from *Bacillus stearothermophilus*. Journal of Dairy Science. 2008;91:1751-1758. DOI: 10.3168/jds.2007/617

[104] Hansson, T, Kaper, T, van der Oost, J, de Vos, W M, Adlercreutz, P. Improved oligosaccharide synthesis by protein engineering of beta-glucosidase CelB from hyperthermophilic *Pyrococcus furiosus*. Biotechnology and Bioengineering. 2001;73:203-210. DOI: 10.1002/bit.1052

[105] Maugard T, Gaunt D, Legoy MD, Besson T. Microwave-assisted synthesis of galacto-oligosaccharides from lactose with immobilized β-galactosidase from *Kluyveromyces lactis*. Biotechnology Letters.2003;25:623-629. DOI: 10.1023/A:1023060030558

[106] Gosling, A, Stevens, G W, Barber, A R, Kentish, S E, Gras, S L. Recent advances refining galactooligosaccharide production from lactose. Food Chemistry. 2010;121:307-318. DOI: 10.1016/j.foodchem.2009.12.063

[107] Huber RE, Kurz G, Wallenfels K. A quantitation of the factors which affect the hydrolase and transgalactosylase activities of beta-galactosidase (*E. coli*) on lactose. Biochemistry.1976;15:1994-2001.

[108] Hsu, C-A, Yu, R-C, Chou, C-C. Purification and characterization of a sodium-stimulated β-galactosidase from *Bifidobacterium longum* CCRC 15708. World Journal of Microbiology and Biotechnology. 2006;22:355-361. DOI 10.1007/s11274-005-9041-0

[109] Seibela, J, Buchholz, K. Tools in Oligosaccharide Synthesis: Current Research and Application. Advances in Carbohydrate Chemistry and Biochemistry. 2010;63:101-138. DOI: 10.1016/S0065-2318(10)63004-1

[110] Kunst, T. Protein Modification to Optimize Functionality: Protein Hydrolysates Galactosidase. In: Whitaker, J R, Voragen, A G J, Wong, D W S, editors. Handbook of Food Enzymology. New York: M. Dekker; 2003, p. 221-236.

[111] Kawamoto, T, Tanaka, A. Entrapment of Biocatalysts by Prepolymer Methods. In: Whitaker, J R, Voragen, A G J, Wong, D W S, editors. Handbook of Food Enzymology. New York: M. Dekker; 2003. p. 331-342.

[112] Urrutia, P, Mateo, C, Guisan, J M, Wilson, L, Illanes, A. Immobilization of *Bacillus circulans* β-galactosidase and its application in the synthesis of galacto-oligosaccharides under repeated-batch operation. Biochemical Engineering Journal. 2013;77:41-48. DOI: 10.1016/j.bej.2013.04.015

[113] Palai, T, Singh, A K, Bhattacharya, P K. Enzyme β-galactosidase immobilized on membrane surface for galacto-oligosaccharides formation from lactose: Kinetic study with feed flow under recirculation loop. Biochemical Engineering Journal. 2014;88:68-76. DOI: 10.1016/j.bej.2014.03.017

[114] Lu. L, Xu, S, Zhao, R, Zhang, D, Li, Z, Li, Y, Xiao, M. Synthesis of galactooligosaccharides by CBD fusion β-galactosidase immobilized on cellulose. Bioresource Technology. 2012;116:327-333. DOI: 10.1016/j.biortech.2012.03.108

[115] Warmerdam, A, Benjamins, E, de Leeuw, T F, Broekhuis, T A, Boom, R M, Janssen, A E M. Galacto-oligosaccharide production with immobilized β-galactosidase in a packed-bed reactor vs. free β-galactosidase in a batch reactor. Food and Bioproducts Processing. 2014;92:383-392. DOI: 10.1016/j.fbp.2013.08.014

[116] Ivanov, A E, Edink, E, Kumar, A, Galaev, I Y, Arendsen, A F, Bruggink, A, Mattiasson, B. Conjugation of penicillin acylase with the reactive copolymer of N-isopropylacrylamide: a step toward a thermosensitive industrial biocatalyst. Biotechnology Progress. 2003;19:1167-1175. DOI: 10.1021/bp0201455

[117] Ward MA, Georgiu TK. Thermoresponsive polymers for biomedical applications. Polymers. 2011;3:1215-1242 DOI: 10.3390/polym3031215

[118] Palai T, Kumar A, Bhattacharya PK. Synthesis and characterization of thermo-responsive poly-N-isopropylacrylamide bioconjugates for application in the formation of galacto-oligosaccharides. Enzyme and Microbial Technology. 2014;55:40-49. DOI: 10.1016/j.enzmitec.2013.12.003

[119] Palai T, Kumar A, Bhattacharya PK. Kinetic studies and model development for the formation ofgalacto-oligosaccharides from lactose using synthesized thermo-responsive bioconjugate. Enzyme and Microbial Technology. 2015;70:42-49. DOI: 10.1016/j.enzmictec.2014.12.010

[120] Torres DPM, Gonçalves MPF, Teixeira JA, Rodrigues L R. Galacto-Oligosaccharides: Production, Properties, Applications, and Significance as Prebiotics. Comprehensive Reviews in Food Science and Food Safety. 2010;9:438-454. DOI: 10.1111/j.1541-4337.2010.00119.x

[121] Guerrero C, Vera C, Novoa C, Dumont J, Acevedo F, Illanes A. Purification of highly concentrated galacto-oligosaccharide preparations by selective fermentation with yeasts. International Dairy Journal.2014;39:78-88. DOI: 10.1016/j.idairyj.2014.05.011

[122] . Venica CI, Bergamini C V, Rebechi S R, Perotti M C Galacto-oligosaccharides formation during manufacture of different varieties of yogurt. Stability through storage LWT - Food Science and Technology. DOI: 10.1016/j.lwt.2015.02.032

[123] Nauta A, Bakker-Zierikzee AM, Schoterman MHC. Galacto-Oligosaccharides. In: Cho, SS, Finocchiaro, ET, editors. Handbook of prebiotics and probiotics ingredients: health benefits and food applications. Boca Raton: Taylor and Francis Group;2010. p. 75-93.

[124] Shoaf K, Mulvey GL, Armstrong GD, Hutkins RW. Prebiotic galactooligosaccharides reduce adherence of enteropathogenic Escherichia coli to tissue culture cells, Infection and. Immunity. 2006;74:6920-6928. DOI: 10.1128/IAI.01030-06

[125] Rowland IR, Tanaka R. The effects of transgalactosylated oligosaccharides on gut flora metabolism in rats associated with a human fecal microflora, Journal of Applied Bacteriology. 1993;74:667-674. DOI: 10.1111/j.1365-2672.1993.tb05201.x

[126] Rivero-Urgell M, Santamaria-Orleans A. Oligosaccharides: application in infant food. Early Human Development. 2001;65:S43-S52. DOI: 10.1016/S0378-3782(01)00202-X

[127] Kim SK, Mendis E. Bioactive compounds from marine processing byproducts – A review. Food Research International. 2006;39:383-393. DOI: 10.1016/j.foodres.2005.10.010

[128] Adrangi S, Faramarsi MA. From bacteria to human: A journey into the world of chitinases. Biotechnology Advances. 2013;31:1786-1795. DOI: 10.1016/j.biotechadv.2013.09.012

[129] Ordonez-Del Pazo T, Antelo LT, Franco-Urıa A, Perez-Martın RI, Sotelo C.G, Alonso AA. Fish discards management in selected Spanish and Portuguese metiers: Identification and potential valorization. Trends in Food Science & Technology. 2014;36:29-43. DOI: 10.1016/j.tifs.2013.12.006

[130] Newton R, Telfer T, Little D. Perspectives on the utilization of aquaculture coproduct in Europe and Asia: Prospects for value addition and improved resource efficiency. Critical Reviews in Food Science and Nutrition. 2014;54:495-510. DOI: 10.1080/10408398.2011.588349

[131] Jung W.-J.; Park R.-D. Bioproduction of Chitooligosaccharides: Present and Perspectives. Marine Drugs. 2014;12:5328-5356. DOI: 10.3390/md12105328

[132] Yang Y, Yu B. Recent advances in the synthesis of chitooligosaccharides and congeners. Tetrahedron. 2014;70:1023-1046. DOI: 10.1016/j.tet.2013.11.064

[133] Mourya VK, Inamdar NN, Choudhari YM. Chitooligosaccharides: Synthesis, Characterization and Applications. Polymer Science, Serie A. 2011;53:583-612. DOI: 10.1134/S0965545X11070066

[134] Cheng, CY, Chang CH, Wu YJ, Li YK. Exploration of Glycosyl Hydrolase Family 75, a Chitosanasefrom *Aspergillus fumigatus*. The Journal of Biological Chemistry. 2006;281:3137-3144. DOI: 10.1074/jbc.M512506200

[135] Fernandes de Assis, C, Araujo, N K, Pagnoncelli, M G B, da Silva Pedrini, M R, Ribeiro de Macedo, G, dos Santos, E S. Chitooligosaccharides enzymatic production by *Metarhizium anisopliae*. Bioprocess and Biosystem Engineering 2010;33:893-899. DOI: 10.1007/s00449-010-0412-z

[136] Gao XA, Zhang YF, Park RD, Huang X, Zhao XY, Xie J, Jin RD. Preparation of chitooligosaccharides from chitosan using crude enzyme of Bacillus cereus D-11. Journal of Applied Biological Chemistry. 2012;55:13-17. DOI: 10.3839/jabc.2011.053

[137] Ming M, Kuroiwa T, Ichikawa S, Sato S, Mukataka S. Production of chitosan-oligosaccharides at high concentration by immobilized chitosanase. Food Science and Technology Research. 2006;12:85-90.

[138] Kuroiwa T, Izuta H, Nabetani H, Nakajima M, Sato S, Mukataka S, Ichikawa S. Selective and stable production of physiologically active chitosan oligosaccharides using an enzymatic membrane bioreactor. Process Biochemistry. 2009;44:283-287. DOI: 10.1016/j.procbio.2008.10.020

[139] No HK, Young PN, Ho LS, Meyers SP. Antibacterial activity of chitosans and chitosan oligomers with different molecular weights. International Journal of Food Microbiology. 2002;74:65-72. DOI: International Journal of Food Microbiology

[140] Li K; Xing R, Liu S, Qin Y, Yu H, Li P. Size and pH effects of chitooligomers on antibacterial activity against *Staphylococcus aureus*. International Journal of Biological Macromolecules. 2014;64:302-305. DOI: 10.1016/j.ijbiomac.2013.11.037

[141] Chung YCh, Su YP, Chen, CC, Jia G, Wang HL, Gaston Wu JC, Lin JG. Relationship between antibacterial activity of chitosan and surface characteristics of cell wall. Acta Pharmacologica Sinica. 2004;25:932-936.

[142] Benhabiles MS, Salah R, Lounici H, Drouiche N, Goosen MFA, Mameri N. Antibacterial activity of chitin, chitosan and its oligomers prepared from shrimp shell waste. Food Hydrocolloids. 2012;29:48-56. DOI: 10.1016/j.foodhyd.2012.02.013

[143] Wang, Y.; Zhou, P.; Yu, J.; Pan, X.; Wang, P.; Lan, W.; Tao, Sh. Antimicrobial effect of chitooligosaccharides produced by chitosanase from Pseudomonas CUY8. Asia Pacific Journal of Clinical Nutrition. 2007;16:174-177.

[144] Ngo DH, Wijesekara I, Vo TS.; Tan QV, Kim SK. Marine food-derived functional ingredients as potential antioxidants in the food industry: An overview. Food Research International. 2011;44:523-529. DOI: 10.1016/j.foodres.2010.12.030

[145] Je JY, Park PJ, Kim SK. Free radical scavenging properties of heterochitooligosaccharides using an ESR spectroscopy. Food and Chemical Toxicology. 2004;42:381-387. DOI: 10.1016/j.fct.2003.10.001

[146] Fernandes JC, Eaton P, Nascimento H, Gião MS, Ramos ÓS, Belo L, Santos-Silva A, Pintado ME, Malcata FX. Antioxidant activity of chitooligosaccharides upon two biological systems: Erythrocytes and bacteriophages. Carbohydrate Polymers. 2010;79:1101-1106. DOI: 10.1016/j.carbpol.2009.10.050

[147] Kim SK, Rajapaksea Ni. Enzymatic production and biological activities of chitosan oligosaccharides (COS): A review. Carbohydrate Polymers. 2005;62:357-368. DOI: 10.1016/j.carbpol.2005.08.012

[148] Halder SK, Adak A, Maity C, Jana A, Das A, Paul T, Ghosh K, Mohapatra PKD, Pati BR, Mondal KC. Exploitation of fermented shrimp-shells hydrolysate as functional food: Assessment of antioxidant, hypocholesterolemic and prebiotic activities. Indian Journal of Experimental Biology. 2013;51:924-934.

[149] Eom TK, Senevirathnea M, Kim SK. Synthesis of phenolic acid conjugated chitooligosaccharides and evaluation of their antioxidant activity. environmental toxicology and pharmacology. 2012;34:519-527. DOI: 10.1016/j.etap.2012.05.004

[150] Lee HW, Park YS, Jung JS, Shin WS. Chitosan oligosaccharides, dp 2–8, have prebiotic effect on the *Bifidobacterium bifidium* and *Lactobacillus* sp. Anaerobe. 2002;8:319-324. DOI: 10.1016/S1075-9964(03)00030-1

[151] Koppová I, Bureš M, Šimůnek J. Intestinal bacterial population of healthy rats during the administration of chitosan and chitooligosaccharides. Folia Microbiologica. 2012;57:295-299. DOI: 10.1007/s12223-012-0129-2

[152] Pan X, Chen F, Wu T, Tang H, Zhao Z. Prebiotic oligosaccharides change the concentrations of short-chain fatty acids and the microbial population of mouse bowel. Journal of Zhejiang University SCIENCE B. 2009;10:258-263.

[153] Gibson, G R, Rastall, R A. Prebiotics: Development & Application. Chichester: John Wiley & Sons Ltd; 2006.

[154] Nguyen, T H, Haltrich, D. Microbial production of prebiotic oligosaccharides. In: McNeil, B, Archer, D, Giavasis, I, Harvey, L. Microbial Production of Food Ingredients, Enzymes and Nutraceuticals. Woodhead Publishing; 2013. p. 494-530. DOI: 10.1533/9780857093547.2.494.ch18.

[155] Zhang, L, Su, Y, Zheng, Y, Jiang, Z, Shi, J, Zhu, Y, Jiang, Y. Sandwich structured enzyme membrane reactor for efficient conversion of maltose into isomaltooligosaccharides. Bioresource Technology. 2010;101:9144-9149. DOI: 10.1016/j.biortech. 2010.07.001

[156] Bharti, S K, Kumar, A, Krishnan, S, Gupta, A K, Kumar, A. Mechanism-based antidiabetic activity of Fructo- and isomalto-oligosaccharides: Validation by *in vivo, in silico* and *in vitro* interaction potential. Process Biochemistry. 2015;50:317-327. DOI: 10.1016/j.procbio.2014.10.014

[157] Qiang, X, YongLie, C, QianBing, W. Health benefit application of functional oligosaccharides. Carbohydrate Polymers. 2009;77:435-441. DOI: 10.1016/j.carbpol.2009.03.016

[158] Meyer, D. Chapter Two - Health Benefits of Prebiotic Fibers. Advances in Food and Nutrition Research. 2015;74:47-91.DOI: 10.1016/bs.afnr.2014.11.002

[159] Candela, M, Maccaferri, S, Turroni, S, Carnevali, P, Brigidi, P. Functional intestinal microbiome, new frontiers in prebiotic design. International Journal of Food Microbiology. 2010;140:93-101. DOI: 10.1016/j.ijfoodmicro.2010.04.017

[160] Vazquez, M J, Alonso, J L, Domınguez, H, Parajo, J C. Xylooligosaccharides: manufacture and applications. Trends in Food Science and Technology. 2000;11:387-393.

[161] Carvalho, A F A, Oliva Neto, P, Silva, D F, Pastore, G M. Xylo-oligosaccharides from lignocellulosic materials: Chemical structure, health benefits and production by chemical and enzymatic hydrolysis. Food Research International. 2013;51:75-85. DOI: 10.1016/j.foodres.2012.11.021

[162] Chung, Y C, Hsu, C K, Ko, C Y, Chan, Y C. Dietary intake of xylooligosaccharides improves the intestinal microbiota, fecal moisture, and pH value in the elderly. Nutrition Research. 2007;27:756-761. DOI: 10.1016/j.nutres.2007.09.014

[163] Moura, P, Cabanas, S, Lourenço, P, Girio, F, Loureiro-Dias, M C, Esteves, M P. *In vitro* fermentation of selected xylo-oligosaccharides by piglet intestinal microbiota. LWT - Food Science and Technology. 2008;41:1952-1961. DOI: 10.1016/j.lwt. 2007.11.007

[164] Veenashri, B R, Muralikrishna, G. *In vitro* anti-oxidant activity of xylo-oligosaccharides derived from cereal and millet brans – A comparative study. Food Chemistry. 2011;126:1475-1481. DOI: 10.1016/j.foodchem.2010.11.163

[165] Broekaert, W F, Courtin, C M, Verbeke, K, Van De Wiele, T, Verstraete, W, Delcour, J A. Prebiotic and other health-related effects of cereal-derived arabinoxylans, arabinoxylan-oligosaccharides, and xylooligosaccharides. Critical Reviews in Food Science and Nutrition. 2011;51:178-194. DOI: 10.1080/10408390903044768

[166] Grootaert, C, Delcour, J A, Courtin, C M, Broekaert, W F, Verstraete, W, Wiele, T V. Microbial metabolism and prebiotic potency of arabinoxylan oligosaccharides in the human intestine. Trends in Food Science and Technology. 2007;18:64-71. DOI: 10.1016/j.tifs.2006.08.004

[167] Fialho, M B, Simões, K, Barros, C A, Bom Pessoni, R A, Braga, M R, Figueiredo-Ribeiro, R C L. Production of 6-kestose by the filamentous fungus *Gliocladium virens* as affected by sucrose concentration. Mycoscience. 2013;54:198-205. DOI: 10.1016/j.myc.2012.09.012

[168] Bekers, M, Laukevics, J, Upite, D, Kaminska, E, Vigants, A, Viesturs, U, Pankova, L, Danilevics, A. Fructooligosaccharide and levan producing activity of *Zymomonas mobilis* extracellular levansucrase. Process Biochemistry. 2002;38:701-706. DOI: 10.1016/S0032-9592(02)00189-9

[169] Álvaro-Benito, M, Abreu, M, Fernández-Arrojo, L, Plou, F J, Jimenez-Barbero, J, Ballesteros, A, Polaina, J, Fernández-Lobato, M. Characterization of a β-fructofuranosidase from *Schwanniomyces occidentalis* with transfructosylating activity yielding the prebiotic 6-kestose. Journal of Biotechnology. 2007;132:75-81. DOI: 10.1016/j.jbiotec.2007.07.939

[170] Marx, S, Winkler, S, Hartmeier, W. Metabolization of beta-(2,6)-linked fructose-oligosaccharides by different bifidobacteria. FEMS Microbiology Letters. 2000;182:163-169. DOI: 10.1111/j.1574-6968.2000.tb09376.x

[171] Porras-Domínguez, J R, Ávila-Fernández, A, Rodríguez-Alegría, M E, Miranda-Molina, A, Escalante, A, González-Cervantes, R, Olvera, C, Munguía, A L. Levan-type FOS production using a *Bacillus licheniformis* endolevanase. Process Biochemistry. 2014;49:783-790. DOI: 10.1016/j.procbio.2014.02.005

[172] Chen, J, Liang, R H, Liu, W, Li, T, Liu, C M, Wu, S S, Wang, Z J. Pectic-oligosaccharides prepared by dynamic high-pressure microfluidization and their *in vitro* fermentation properties. Carbohydrate Polymers. 2013;91:175-182. DOI: 10.1016/j.carbpol.2012.08.021

[173] Gullon, B, Gomez, B, Martinez-Sabajanes, M, Yanez, R, Parajo, J C, Alonso, J L. Pectic oligosaccharides: Manufacture and functional properties. Trends in Food Science and Technology. 2013;30:153-161. DOI: 10.1016/j.tifs.2013.01.006

[174] Gutiérrez-Román MI, Dunn MF, Tinoco-Valencia R, Holguín-Meléndez F, Huerta-Palacios G, Guillén-Navarro K. Potentiation of the synergistic activities of chitinases ChiA, ChiB and ChiC from Serratia marcescens CFFSUR-B2 by chitobiase (Chb) and

chitin binding protein (CBP). World Journal of Microbiology Biotechnology. 2014;30:33-42. DOI: 10.1007/s11274-013-1421-2

[175] Karan R, Cape MD, DasSarma S. Function and biotechnology of extremophilic enzymes in low water activity. Aquatic Biosystems. 2012;8:4.

[176] Pinelo M, Jonsson G, Meyer AS. Membrane technology for purification of enzymatically produced oligosaccharides: Molecular and operational features affecting performance. Separation and Purification Technology. 2009;70:1-11. DOI:10.1016/j.seppur.2009.08.010

[177] Cui, S W. Food Carbohydrates: Chemistry, Physical Properties, and Applications. CRC Press - Taylor & Francis Group. 2005.

[178] Jakob, F, Pfaff, A, Novoa-Carballal, R, Rübsamc, H, Becker, T, Vogel, R F. Structural analysis of fructans produced by acetic acid bacteria reveals a relation to hydrocolloid function. Carbohydrate Polymers. 2013;92:1234-1242. DOI: 10.1016/j.carbpol.2012.10.054

[179] Bruzzese, E, Volpicelli, M, Squeglia, V, Bruzzese, D, Salvini, F, Bisceglia, M, Lionetti, P, Cinquetti, M, Iacono, G, Amarri, S, Guarino, A. A formula containing galacto- and fructo-oligosaccharides prevents intestinal and extra-intestinal infections: An observational study. Clinical Nutrition. 2009;28:156-161. DOI: 10.1016/j.clnu.2009.01.008

[180] Gomes, A M P, Malcata, F X. *Bifidobacterium* spp. and *Lactobacillus acidophilus*: biological, biochemical, technological and therapeutical properties relevant for use as probiotics. Trends in Food Science and Technology. 1999;10:139-157. DOI: 10.1016/S0924-2244(99)00033-3

[181] Yen, C H, Kuo, Y W, Tseng, Y H, Lee, M C, Chen, H L. Beneficial effects of fructooligosaccharides supplementation on fecal bifidobacteria and index of peroxidation status in constipated nursing-home residents: A placebo-controlled, diet-controlled trial. Nutrition. 2011;27:323-328. DOI: 10.1016/j.nut.2010.02.009

[182] Jakobsdottir, G, Nyman, M, Fak, F. Designing future prebiotic fiber to target metabolic syndrome. Nutrition. 2014;30:497-502. DOI: 10.1016/j.nut.2013.08.013

[183] Choque Delgado, G T, Tamashiro, W M S C, Pastore, G M. Immunomodulatory effects of fructans. Food Research International. 2010;43:1231-1236. DOI: 10.1016/j.foodres.2010.04.023

[184] Juskiewicz, J, Semaskaite, A, Zdunczyk, Z, Wroblewska, M, Gruzauskas, R, Juskiewicz, M. Minor effect of the dietary combination of probiotic *Pediococcus acidilactic*i with fructooligosaccharides or polysaccharidases on beneficial changes in the cecum of rats. Nutrition Research. 2007;27:133-139. DOI: 10.1016/j.nutres.2007.01.005

[185] Licht, T R, Ebersbach, T, Frøkiær, H. Prebiotics for prevention of gut infections. Trends in Food Science and Technology. 2012;23:70-82. DOI: 10.1016/j.tifs.2011.08.011

[186] Van den Heuvel, E G H M, Muijs, T, Brouns, F, Hendriks, H F J. Short-chain fructo-oligosaccharides improve magnesium absorption in adolescent girls with a low calcium intake. Nutrition Research. 2009;29:229-237. DOI: 10.1016/j.nutres.2009.03.005

[187] Losada, M A, Olleros, T. Towards a healthier diet for the colon: the influence of fructooligosaccharides and lactobacilli on intestinal health. Nutrition Research. 2002;22:71-84. DOI: 10.1016/S0271-5317(01)00395-5

[188] Niness, K R. Inulin and Oligofructose: What Are They? The Journal of Nutrition. 1999; Supplement:1402S-1406S.

[189] Glibowski, P, Pikus, S, Jurek, J, Kotowoda, M. Factors affecting inulin crystallization after its complete dissolution. Carbohydrate Polymers. 2014;110:107-112. DOI: 10.1016/j.carbpol.2014.03.080

[190] Morris C, Morris GA. The effect of inulin and fructo-oligosaccharide supplementation on the textural, rheological and sensory properties of bread and their role in weight management: A review. Food Chemistry. 2012;133:237-248. DOI: 10.1016/j.foodchem.2012.01.027

[191] Beserra, B T S, Fernandes, R, Rosario, V A, Mocellin, M C, Kuntz, M G F, Trindade, E B S M. A systematic review and meta-analysis of the prebiotics and synbiotics effects on glycaemia, insulin concentrations and lipid parameters in adult patients with overweight or obesity. Clinical Nutrition. DOI: 10.1016/ j.clnu.2014.10.004

[192] Karimi, R, Azizi, M H, Ghasemlou, M, Vaziri, M. Application of inulin in cheese as prebiotic, fat replacer and texturizer: A review. Carbohydrate Polymers. DOI: 10.1016/j.carbpol.2014.11.029

[193] Oliveira, R P S, Perego, P, Oliveira, M N, Converti, A. Growth, organic acids profile and sugar metabolism of *Bifidobacterium lactis* in co-culture with *Streptococcus thermophilus*: The inulin effect. Food Research International. 2012;48:21-27. DOI: 10.1016/j.foodres.2012.02.012

[194] Dehghan, P, Gargari, B P, Jafar-abadi, M A. Oligofructose-enriched inulin improves some inflammatory markers and metabolic endotoxemia in women with type 2 diabetes mellitus: A randomized controlled clinical trial. Nutrition. 2014;30:418-423. DOI: 10.1016/j.nut.2013.09.005

[195] Apolinário, A C, Damasceno, B P G L, Beltrão, N E M, Pessoa, A, Converti, A, Silva, J A. Inulin-type fructans: A review on different aspects of biochemical and pharmaceutical technology. Carbohydrate Polymers. 2014;101:368-378. DOI: 10.1016/j.carbpol.2013.09.081

[196] Adebola, O, Corcoran, O, Morgan, W A. Protective effects of prebiotics inulin and lactulose from cytotoxicity and genotoxicity in human colon adenocarcinoma cells. Food Research International. 2013;52:269-274. DOI: 10.1016/j.foodres.2013.03.024

[197] Martínez-Villaluenga, C, Cardelle-Cobas, A, Corzo, N, Olano, A. Study of galactooligosaccharide composition in commercial fermented milks. Journal of Food Composition and Analysis. 2008;21:540-544. DOI:10.1016/j.jfca.2008.05.008

[198] Crittenden, R G, Playne, M J. Production, properties and application of food grade oligosaccharides. Trends in Food Science & Technology. 1996;71:353-361.

[199] Sarabia-Sainz, H M, Armenta-Ruiz, C, Sarabia-Sainz, J A, Guzmán-Partida, A M, Ledesma-Osuna, A I, Vázquez-Moreno, L, Montfort, G R C. Adhesion of enterotoxigenic *Escherichia coli* strains to neoglycans synthesised with prebiotic galactooligosaccharides. Food Chemistry. 2013;141:2727-2734. DOI:10.1016/j.foodchem.2013.05.040

[200] Puccio, G, Cajozzo, C, Meli, F, Rochat, F, Grathwohl, D, Steenhout, P. Clinical evaluation of a new starter formula for infants containing live *Bifidobacterium longum* BL999 and prebiotics. Nutrition. 2007;23:1-8. DOI:10.1016/j.nut.2006.09.007

[201] Zhong, Y, Cai, D, Cai, W, Geng, S, Chen, L, Han, T. Protective effect of galactooligosaccharide-supplemented enteral nutrition on intestinal barrier function in rats with severe acute pancreatitis. Clinical Nutrition. 2009;28:575-580. DOI:10.1016/j.clnu.2009.04.026

[202] Sangwan, V, Tomar, S K, Ali, B, Singh, R R B, Singh, A K. Galactooligosaccharides reduce infection caused by *Listeria monocytogenes* and modulate IgG and IgA levels in mice. International Dairy Journal. 2015;41:58-63. DOI:10.1016/j.idairyj.2014.09.010

[203] Hernandez, O, Ruiz-Matute, A I, Olano, A, Moreno, F J, Sanz, M L. Comparison of fractionation techniques to obtain prebiotic galactooligosaccharides. International Dairy Journal. 2009;19:531-536. DOI:10.1016/j.idairyj.2009.03.002

[204] Frenzel, M, Zerge, K, Clawin-Radecker, I, Lorenzen, P C. Comparison of the galacto-oligosaccharide forming activity of different β-galactosidases. LWT - Food Science and Technology. 2015;60:1068-1071. DOI:10.1016/j.lwt.2014.10.064

[205] Tymczyszyn, E E, Sosa, N, Gerbino, E, Hugo, A, Gómez-Zavaglia, A, Schebor, C. Effect of physical properties on the stability of *Lactobacillus bulgaricus* in a freeze-dried galacto-oligosaccharides matrix. International Journal of Food Microbiology. 2012;155:217-221. DOI:10.1016/j.ijfoodmicro.2012.02.008

[206] Bruno-Barcena, J M, Azcarate-Peril, M A. Galacto-oligosaccharides and colorectal cancer: Feeding our intestinal probiome. Journal of Functional Foods. 2015;12:92-108. DOI:10.1016/j.jff.2014.10.029

[207] Torres, D P M, Bastos, M, Gonçalves, M P F, Teixeira, J A, Rodrigues, L R. Water sorption and plasticization of an amorphous galacto-oligosaccharide mixture. Carbohydrate Polymers. 2011;83:831-835. DOI: 10.1111/j.1541-4337.2010.00119.x

[208] Adebola, O, Corcoran, O, Morgan, W,A. Protective effects of prebiotics inulin and lactulose from cytotoxicity and genotoxicity in human colon adenocarcinoma cells. Food Research International. 2013;52:269-274. DOI:10.1016/j.foodres.2013.03.024

[209] Venema, K. Intestinal fermentation of lactose and prebiotic lactose derivatives, including human milk oligosaccharides. International Dairy Journal. 2012;22:123-140. DOI:10.1016/j.idairyj.2011.10.011

[210] Mayer, J., Kranz, B., Fischer, L. Continuous production of lactulose by immobilized thermostable -glycosidase from *Pyrococcus furiosus*. Journal of Biotechnology. 2010;145:387-393. DOI:10.1016/j.jbiotec.2009.12.017

[211] Schuster-Wolff-Bühring, R, Fischer, L, Hinrichs, J. Production and physiological action of the disaccharide lactulose. International Dairy Journal. 2010;20:731-741. DOI: 10.1016/j.idairyj.2010.05.004

[212] Seki, N, Saito, H. Lactose as a source for lactulose and other functional lactose derivatives. International Dairy Journal. 2012;22:110-115. DOI:10.1016/j.idairyj.2011.09.016.

[213] Muzzarelli, R A A, Boudrant, J, Meyer, D, Manno, N, DeMarchis, M, Paoletti, M G. Current views on fungal chitin/chitosan, human chitinases, food preservation, glucans, pectins and inulin: A tribute to Henri Braconnot, precursor of the carbohydrate polymers science, on the chitin bicentennial. Carbohydrate Polymers. 2012;87:995-1012. DOI:10.1016/j.carbpol.2011.09.063

[214] Prashanth, K V H, Tharanathan, R N. Chitin/chitosan: modifications and their unlimited application potential: an overview. Trends in Food Science & Technology. 2007;18:117-131. DOI:10.1016/j.tifs.2006.10.022

[215] Chung, Y C, Hsub, C K, Koa, C Y, Chana, Y C. Dietary intake of xylooligosaccharides improves the intestinal microbiota, fecal moisture, and pH value in the elderly. Nutrition Research. 2007;27:756-761. DOI:10.1016/j.nutres.2007.09.014

[216] Carvalho, A F A, Oliva Neto, P, Silva, D,F, Pastore, G,M. Xylo-oligosaccharides from lignocellulosic materials: Chemical structure, health benefits and production by chemical and enzymatic hydrolysis. Food Research International. 2013;51:75-85. DOI: 10.1016/j.foodres.2012.11.021

[217] Veenashri, B R, Muralikrishna, G. *In vitro* anti-oxidant activity of xylo-oligosaccharides derived from cereal and millet brans – A comparative study. Food Chemistry. 2011;126:1475-1481. DOI:10.1016/j.foodchem.2010.11.163

[218] Bharti, S,K, Kumar, A, Krishnan, S, Gupta, A K, Kumar A. Mechanism-based antidiabetic activity of Fructo- and isomalto-oligosaccharides: Validation by *in vivo, in silico* and *in vitro* interaction potential. Process Biochemistry. 2015;50:317-327. DOI:10.1016/j.procbio.2014.10.014

[219] Candela, M, Maccaferri, S, Turroni, S, Carnevali, P, Brigidi, P. Functional intestinal microbiome, new frontiers in prebiotic design. International Journal of Food Microbiology. 2010;140:93-101. DOI:10.1016/j.ijfoodmicro.2010.04.017

[220] Nguyen, T H, Haltrich, D. Microbial production of prebiotic oligosaccharides. In: McNeil B, Archer D, Giavasis I, Harvey L. Microbial Production of Food Ingredients, Enzymes and Nutraceuticals. Woodhead Publishing; 2013. p. 494-530. DOI: 10.1533/9780857093547.2.494. ch18.

[221] Chen, H L, Lu, Y H, Lin, J, Ko, L I. Effects of fructooligosaccharide on bowel function and indicators of nutritional status in constipated elderly men. Nutrition Research. 2000;20:1725-1733.

[222] Cheng, W T, Lin, S Y. Processes of dehydration and rehydration of raffinose pentahydrate investigated by thermal analysis and FT-IR/DSC microscopic system. Carbohydrate Polymers. 2006;64:212-217. DOI:10.1016/j.carbpol.2005.11.024

[223] Huebner, J., Wehling, R L, Hutkins, R W. Functional activity of commercial prebiotics. International Dairy Journal. 2007;17:770-775. DOI:10.1016/j.idairyj.2006.10.006

[224] Anthony, J C, Merriman, T N, Heimbach, J T. 90-Day oral (gavage) study in rats with galactooligosaccharides syrup. Food and Chemical Toxicology. 2006;44:819-826. DOI: 10.1016/j.fct.2005.10.012

[225] Osman, A, Tzortzis, G, Rastall, R A, Charalampopoulos, D. A comprehensive investigation of the synthesis of prebiotic galactooligosaccharides by whole cells of *Bifidobacterium bifidum* NCIMB 41171. Journal of Biotechnology. 2010;150:140-148. DOI: 10.1016/j.jbiotec.2010.08.008

[226] Davis, L M G, Martínez, I, Walter, J, Hutkins R. A dose dependent impact of prebiotic galactooligosaccharides on the intestinal microbiota of healthy adults. International Journal of Food Microbiology. 2010;144:285-292. DOI:10.1111/jam.12415

[227] Goulas, A, Tzortzis, G, Gibson, G R. Development of a process for the production and purification of α- and β-galactooligosaccharides from *Bifidobacterium bifidum* NCIMB 41171. International Dairy Journal. 2007;17:648-656. DOI:10.1016/j.idairyj.2006.08.010

[228] Searle, L E J, Jones, G, Tzortzis, G, Woodward, M J, Rastall, R A, Gibson, G R, La Ragione, R M. Low molecular weight fractions of BiMuno exert immunostimulatory properties in murine macrophages. Journal of Functional Foods. 2012;4:941-953. DOI: 10.1016/j.jff.2012.07.002

[229] Förster-Fromme, K, Schuster-Wolff-Bühring, R, Hartwig, A, Holder, A, Schwiertz, A, Bischoff, S C, Hinrichs J. A new enzymatically produced 1-lactulose: A pilot study to test the bifidogenic effects. International Dairy Journal. 2011;21:940-948. DOI:10.1016/j.idairyj.2011.07.002

[230] Shen, Q, Yang, R, Hua, X, Ye, F, Wang, H, Zhao, W, Wang K. Enzymatic synthesis and identification of oligosaccharides obtained by transgalactosylation of lactose in

the presence of fructose using b-galactosidase from *Kluyveromyces lactis*. Food Chemistry. 2012;135:1547-1554. DOI:10.1016/j.foodchem.2012.05.115

[231] Santos, M I, Gerbino, E, Araujo-Andrade, C, Tymczyszyn, E E, Gómez-Zavaglia, A. Stability of freeze-dried *Lactobacillus delbrueckii* subsp. *bulgaricus* in the presence of galacto-oligosaccharides and lactulose as determined by near infrared spectroscopy. Food Research International. 2014;59:53-60. DOI:10.1016/j.foodres.2014.01.054

[232] Dilokpimol, A, Nakai, H, Gotfredsen, C H, Appeldoorn, M, Baumann, M J, Nakai, N, Schols H A, Hachem, M A, Svensson B. Enzymatic synthesis of b-xylosyl-oligosaccharides by transxylosylation using two b-xylosidases of glycoside hydrolase family 3 from *Aspergillus nidulans* FGSC A4. Carbohydrate Research. 2011;346:421-429. DOI: 10.1016/j.carres.2010.12.010

[233] Aam, B B, Heggset, E B, Norberg, A L, Sørlie, M, Vårum, K M, Eijsink, V G H. Production of Chitooligosaccharides and Their Potential Applications in Medicine. Marine Drugs. 2010;8:1482-1517. DOI:10.3390/md8051482

[234] Jeon, Y J, Park, P J, Kim, S K. Antimicrobial effect of chitooligosaccharides produced by bioreactor. Carbohydrate Polymers. 2001;44:71-76.

[235] Jeon, Y J, Kim, S K. Production of chitooligosaccharides using an ultrafiltration membrane reactor and their antibacterial activity. Carbohydrate Polymers. 2000;41:133-141.

[236] Kim, S K. Chitin, Chitosan, Oligosaccharides and Their Derivatives - Biological Activities and Applications. CRC Press; 2011.

[237] Kim, Y M, Kang, H K, Moon, Y H, Nguyen, T T H, Day, D F, Kim, D. Production and Bioactivity of Glucooligosaccharides and Glucosides Synthesized using Glucansucrases. In: Moreno, F J, Sanz, M L, editors. Food Oligosaccharides - Production, Analysis and Bioactivity. Wiley Blackwell, IFT Press; 2014. ch10.

[238] Vardakou, M, Palop, C N, Christakopoulos, P, Faulds, C B, Gasson, M A, Narbad A. Evaluation of the prebiotic properties of wheat arabinoxylan fractions and induction of hydrolase activity in gut microflora. International Journal of Food Microbiology. 2008;123:166-170. DOI:10.1016/j.ijfoodmicro.2007.11.007

[239] Wicker, L, Kim, Y, Kim, M J, Thirkield, B, Lin, Z, Jung, J. Pectin as a bioactive polysaccharide e Extracting tailored function from less. Trends in Food Science & Technology. 2014;42:251-259. DOI:10.1016/j.foodhyd.2014.01.002

[240] Holck, J, Hjernø, K, Lorentzen, A, Vigsnæs, L K, Hemmingsen, L, Licht, T R, Mikkelsena, J D, Meyer, A S. Tailored enzymatic production of oligosaccharides from sugar beet pectin and evidence of differential effects of a single DP chain length difference on human faecal microbiota composition after *in vitro* fermentation. Process Biochemistry. 2011;46:1039-1049. DOI:10.1016/j.procbio.2011.01.013

[241] Rastall, R A, Gibson, G R. Recent developments in prebiotics to selectively impact beneficial microbes and promote intestinal health. Current Opinion in Biotechnology. 2015,32:42-46. DOI:10.1016/j.copbio.2014.11.002

Chapter 3

New Cheese-Like Food Production from Soy Milk — Utility of Soy Milk Curdling Yeast

Makoto Kanauchi, Sakiko Hatanaka and
Makoto Shimoyamada

Additional information is available at the end of the chapter

http://dx.doi.org/10.5772/60848

Abstract

Soybeans are a traditional food in eastern Asia, particularly in Japan and China. They were eaten in 100 BC in China. The beans can be processed into *Tofu*, soy milk, fermented seasonings, soy sauce or *Miso* paste, and *Natto* and green beans. Soybeans have rich nutrition, protein lipid, and other functional substances such as isoflavones. However, soybeans are difficult to process for use as food because of tissue and cell wall hardness. Therefore, soybeans are conducted to do some treatments, *e.g.*, boiling, steaming, roasting, crushing/grinding, and some enzyme treating, to eat soy protein easily. Soy storage proteins mainly comprise two proteins as 7S globulin composed with β-conglycinin and 11S globulin containing glycinin composed of 5 subunits. β-Conglycinin, included in 7S globulin, is composed of three subunits.

To modify the physical properties of soy protein, a new type of enzyme for curdling soybean milk enzyme was purified as an extract from yeast. Yeast producing curdling soybean milk enzyme, the SCY003 strain, was isolated from 1345 yeast strains. According to the morphology, physiology, and molecular and characteristics, SCY003 was identified as *Saccharomyces bayanus*. The soy milk curdling enzyme having proteolytic activity was approximately 45 kDa and monomer protein. The optimum pH for the protease activity was pH 7.5; the optimum temperature was 50°C.

The enzyme cleaved the β-conglycinin as α–, α'-, and part of glycinin as A_3 A_4, A_{1b}, and A_2 in soy protein by endoproteolysis. Soybean protein became loosely curdled with the addition of other proteases from microorganisms or plants. Soybean milk curdled after cleaving endoproteolysis enzyme in SCY003 strain.

The breaking point of curd curdled by enzyme was 58.4% strain. Their breaking stress was 10,900 ($N \cdot m^{-2}$). The brittleness point is 81.2% and 10,200 ($N \cdot m^{-2}$), and the brittleness of this curd produced using the enzyme was 727 ($N \cdot m^{-2}$). Brittleness of the curd produced by the enzyme was less. Their breaking point was greater than that of the curd produced using the glucono-δ-lactone (GDL). Furthermore, the curd had sticky and chewy texture. The curd made by enzyme has resilience more than normal *Tofu*. It is considered that the curd produced by enzyme was not like *Tofu* rheologically.

Keywords: Soybean milk, curdling, *Saccharomyces bayanus*, protease

1. Introduction

1.1. Utility of soybeans

Soybeans have been used as traditional foods from ancient times. They are rich in nutrients such as rich proteins, lipids, and others. Furthermore, soybeans can be eaten after processing in various ways.

Eastern Asian people and particularly Japanese people have eaten soybeans after various stages of processing. Soybean seedlings are eaten in many dishes as bean sprouts. Furthermore, soybeans are eaten as green beans in the pod after boiling, as *Edamame*. Soybean flour made from roasted soybeans is eaten as *Kinako* powder. After boiling soybeans, the beans can be fermented can using molds to produce *Chi* or *Tempe*. Furthermore, soy sauce is made from a molded mixture with boiled beans and roasted wheat, and salted water. Soy paste is made from fermented boiled soybeans and *Koji* with salt. The resultant umami taste is an extract from the bean, facilitated by an enzyme reaction because it is thought that umami components are stored as proteins in hard tissue.

After hard tissues in boiled soybeans are crushed and ground physically, the soluble fraction is extracted as soy milk. Soy milk is processed as *Yuba* from a soybean sheet, and *Tofu* is produced. Regarding the insoluble fraction, spent soy is also eaten as *Okara*. Finally, compressed soybeans produce oil that is widely used as cooking oil. The residue of oil pressing can then be used for soy sauce production or soy protein for food manufacture. Comparing soy products to milk casein, which is eaten as cheese, soy products are not used as widely as food. One reason is that soybean curd such as *Tofu* lacks taste and has less elastic properties and texture compared to cheese. Therefore, to make rich nutrition and produce a food that has good taste and texture, soybean protein is modified by enzymes and is extracted by microorganisms [1].

1.2. History of soybeans

Theories about the origin of soybean use and cultivation remain controversial in their details. By some accounts, soybeans used as food originated in the area of Manchuria in China and

http://dx.doi.org/10.5772/60848

Siberia in Russia. Alternatively, their use as food originated in southern China [1]. Yet another possible history holds that soybeans were bred from wild soybeans as *Glycine soja* in China. In fact, the bean was present in ancient Japan: beans were found in the bottom of an earthen vessel produced in the middle Jomon period (3000–2000 B.C.) in Japan. However, soybeans have not been recorded as eaten in that period.

In China, the first literal record can be found in a Chinese dictionary published in 100 B.C. The dictionary inserted "*Chi*," representing fermenting soybeans with salt. Furthermore, archaic *Miso* paste and soy sauce fermented by soybeans were described in the Chinese text "Qi-min-yao-shu," published in the sixth century.

In Japan also, "*Chi*" fermented soybeans with salt was recorded in the Taiho Code in A.D. 701. It is considered that soybean fermentation practices diffused from China. Tofu was recorded in the tenth century in China. It was recorded in the same period in Japan. In Japan, it was initially eaten as a vegetarian dish for Buddhists because, during the Kamakura period, people gradually became more and more vegetarian in their eating practices. In the Edo Era, a *Tofu* recipe book was published, with 283 recipes explained in it.

1.3. Consumption worldwide and in Japan

Soybeans were produced only in eastern Asia for a long time. In contrast, other cereals such as rice, barley, wheat, and corn have diffused throughout the world. Moreover, it is considered that some other endemic bean or pea or pulse had become cultivated in each area already [1].

For instance, in central Asia, broad beans were cultivated, as were chick peas in India, shell peas in western Asia, and kidney beans and ground peas in North America. Nevertheless, soybeans have been cultivated in the United States as oil seed crops since the 1920s. Furthermore, the crop has begun to be cultivated in Canada and South America. In 2012, approximately 82 million tons of soybeans were harvested, with the United States accounting for 34% of the world production. Brazil harvested approximately 66 million tons that year, accounting for 27% of world production. Argentina harvested approximately 40 million tons, or 17% of world production. Therefore, most soybeans (over 80%) consumed worldwide are now produced and harvested in North and South America [2] (Table 1).

In Japan, soybean production was sufficient to provide for domestic consumption until the Taisho Era [1]. Soybean consumption in Japan has been high, but it decreased after the Taisho Era. In 2013, 30 million tons were consumed, but only 240 thousand tons were harvested domestically. That figure is less than 0.1% of the world production amount. Soybeans used domestically account for 104 thousand tons for feed, 6 thousand tons for seed, and 1.9 million tons for oilseeds, all together accounting for 70% of the 30 million tons consumed. Furthermore, those figures indicate that only 30% of soybeans are used as food. The self-sufficiency ratio of soybeans was 97% in 1947. It decreased gradually to 28% in 1959, 11% in 1965, and 7% in 2013 [3]. The ratios of soybeans used for food are 49% used for *Tofu*, 13% used for *Miso* paste and *Natto*, 4% used for soy milk, and 3.5% used for soy sauce production. As mentioned earlier, some soy sauce production companies have used soy meal after oil pressing to produce soy sauce (Table 2).

	Production 2012 (10 thousand tons)	Share of world production (%)
United States	8205	33.9
Brazil	6585	27.2
Argentina	4010	16.6
India	1467	6.1
China	1305	5.4
Canada	509	2.1
Paraguay	434	1.8
Uruguay	300	1.2
Ukraine	241	1.0
Bolivia	206	0.9
Russia	181	0.7
Indonesia	84	0.3
South Africa	65	0.3
Nigeria	58	0.2
North Korea	35	0.1
Japan	24	0.1
Myanmar	21	0.1
Others	454	1.9
World total	24,184	100

Source: http://www.maff.go.jp/j/seisan/ryutu/daizu/d_data/pdf/014.pdf

Table 1. Soybeans production

	2003	2013
Miso paste	138	123
Soy sauce	38	33
Tofu and fried tofu	494	454
Natto	137	125
Frozen *tofu*	30	20
Soy milk	19	40
Delicatessen of soybean	33	30
Kinako soy powder	17	18
Other	128	93
Total	1034	936

(Unit: thousand tons)

Ref. http://www.maff.go.jp/j/seisan/ryutu/daizu/d_data/

Table 2. Changes in amount of soybeans for applications

2. Nutrition

2.1. Soybean protein

Soybeans contain 35% protein as storage protein, which is used for nutrition during germination. That storage protein is stored in granules, called protein bodies, of about 5–8 μm diameter. Soluble soy protein is extracted from insoluble protein bodies that are burst during soy milk and *Tofu* production. All soy protein is stored in the protein body [1]. Other proteins exist as nonstorage proteins, containing important physiological proteins such as trypsin inhibitors. When an animal ingests a trypsin inhibitor, a digestive enzyme, trypsin activity is inhibited by combination of the trypsin inhibitor specifically with trypsin. Consequently, the pancreas works excessively to secrete and supplement trypsin activity [1]. However, after heating, the inhibitor loses its inhibitory activity and does not bind with trypsin.

Protein digestibility-corrected amino acid score (PDCAAS) values are used to evaluate the protein quality based on the amino acid requirements of humans and their ability to digest it. It was long thought that the amount of amino acid requirements of humans dictate a low score for soybeans because methionine and cysteine residues, sulfur amino acids, in soybean storage proteins have a low composition. The score was only 86 points based on the amino acid requirements of a developing rat. However, in 1985, the score was modified to 100 points, the same as milk and eggs, based on the amino acid requirements of humans. Soy storage proteins are rich in nutrition for human needs [4].

Throughout the world, the recently improving healthy image of soy protein is interesting. In particular, the health benefits of soy foods attract attention in the United States. Health claims are authorized by the Food and Drug Administration (FDA) in the United States: foods containing 6.25 g of soy protein or more can be said by manufacturers to reduce the risk of heart disease if a consumer ingests 25 g/day of soy protein [5]. In Japan, some soybean containing foods are manufactured as *Tokuho*: government-approved foods for specified health benefits, as for hypocholesterolemic activity in this case.

The taxonomy of soybean storage protein has been conducted according to the sedimentation coefficient by an ultracentrifugal fraction as 2S globulin, 7S globulin, 11S globulin, and 15S globulin. Yamauchi [1] reported details of soybean proteins: 2S globulin contained α-conglycinin, 7S globulin composed with β-conglycinin and γ-conglycinin, and 11S globulin-containing glycinin. In addition, the 11S globulin composes hexamer. It is a 350,000 Da protein. Furthermore, their proteins are composed with five subunits as G_1–G_5; their subunit was 10 polypeptides as A_{1a}, A_2, A_{1b}, A_3, A_5A_4, A_4, B_2, B_{1b}, B_4, and B_3. Their polypeptides are combined specifically as $A_{1a}B_{1b}$, $A_{1b}B_2$, A_2B_{1a}, A_3B_4, and $A_5A_4B_3$. In fact, β-conglycinin, called 7S globulin, combines a dimer protein and a monomer protein, which are 150,000–200,000 Da, or an average of 180,000 Da. They are composed of three subunits: an α-subunit of 63,000 Da, an α'-subunit of 67,000 Da, and a β-subunit of 48,000 Da. The protein has a low concentration of sulfur amino acids. In particular, the β-subunit does not contain methionine, cysteine, and tryptophan [6].

2.2. Lipid

Soybeans have 20% lipids. The lipid concentration varies among harvested regions. Soybeans harvested in the United States have more lipids than those in China [1]. A main reason is that soybeans there have long been bred and modified to contain high oil concentrations as oilseed.

	Palmitic acid	Stearic acid	Oleic acid	Linoleic acid	Linolenic acid
Pork fat	26.2	13.5	42.9	9	0.3
Beef fat	38.7	3.8	42.1	2.3	2.4
Milk lipid	31.1	9.2	21.7	1.6	0.4
Soybean oil	10.5	3.2	22.3	54.5	8.3
Sunflower oil	trace	4	27.6	58.3	trace
Cotton oil	27.3	3.1	16.7	50.4	trace
Safflower oil	6	3.4	12.2	77	0.3

Source: Yamauchi and Ookubo (1992).

Table 3. Components of fatty acids in foods (%)

Components of fatty acids in foods shows Table 3. Soy oil comprises a small amount of saturated fatty acids, such as palmitic acid and stearic acid, and large amount of unsaturated fatty acids such as oleic acid, linoleic acid, and linolenic acid. Polyunsaturated fatty acids (PUFAs) containing more than unsaturated bonds are important nutrition as necessary lipids for humans. Soybeans have over 60% PUFA. In particular, one kind of PUFA as linoleic acid contained approximately 54.5%.

Actually, PUFAs in animal lipids have low concentration. Therefore, they are insufficient nutritionally. Saturated fats and unsaturated fats are ideally in the following ratio: saturated–unsaturated (1:2) [1]. Soybean lipids were well known to be much stable against oxidation because they are covered as oil body particle by oleosin and other proteins.

2.3. Isoflavone

Isoflavone is one kind of flavonoid (Fig. 1). Fabaceae sp. contain high concentrations (Fig. 1).

Generally, soybeans have totally 12 isoflavones in 3 aglycones, and they have three types of glycosides as glucoside, acetyl-glycoside and malonyl-glycosides: genistein, daidzein, glystein, genistin, daidzin, glycitin, acetyl-genistin, acetyl-daidzin, acetyl-glycitin, malonyl-genistin, malonyl-daidzin and malonyl-glycitin [7]. After soybean consumption, glycoside isoflavone, which is contained in food as soy milk or *Tofu*, hydrolyzes aglycon and glycoside by bacteria in intestines. Their aglycon are absorbed by the body. Genistein and daidzein have estrogenic effects and hormone-like activity. The isoflavone binding with estrogen receptor

http://dx.doi.org/10.5772/60848

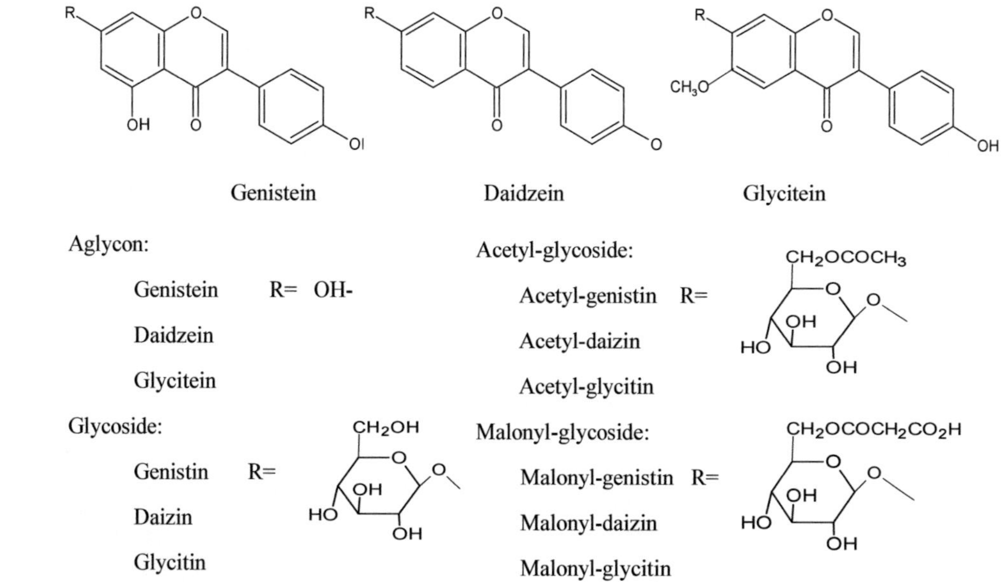

Figure 1. Isoflavone structure.

reacts as an estrogen agonist in the human body. Their substances from plants having estrogenic effects are called "plant estrogen." *Miso* paste has a high level of isoflavone because glucosyl isoflavone is hydrolyzed to their related aglycons.

Many Japanese and throughout eastern Asia intake isoflavone from soybeans. Some researchers have reported negative opinions about plant estrogen [8]. Isoflavones are produced via phenyl–propanoid pathway from phenylalanine in plants. Two intermediate substances, naringenins, are converted to genistein by two specific enzymes in soybeans: isoflavone synthetase and dehydrogenase. Chalcones are converted to daidzein by three specific enzymes in soybean: chalcone reductase, chalcone isomerase, and isoflavone synthetase. Isoflavone and a similar substance, phytoalexin, are used as antibacterial substances against phytopathogenic fungi and bacteria. In addition, they grow well as root nodule bacteria at the root for nitrogen fixation [9, 10].

Other beans and peas, legumes, have isoflavones: chickpeas have biochanin A [11]; alfalfa has formononetin and coumestrol [12]; and ground peas have genistein [13].

Isoflavones in many plants store glucosyl, malonyl-glucosyl, and acetyl-glucosyl conjugate as hydrophilic substances. After invasion of phytopathogenic fungus and bacteria, the glucosyl conjugate isoflavone is transferred to infested wounds, where it hydrolyzes for phytoalexin [14]. Isoflavone has health functions against climacteric disturbances and type 2 diabetes. In Japan, soybean isoflavones are a *Tokuho* (government-approved food for specified health purpose) for the prevention of osteoporosis. The Ministry of Health, Labour and Welfare in Japan alerts consumers to avoid overdosing on isoflavones. The amount of isoflavone intake was 30 mg a day, omitting isoflavone intake from meals and by supplements.

3. Food processing

3.1. Formation of *Tofu* curdling

Soy proteins have properties that produce curd to add specific metal ions. The property is applied for *Tofu* production. Tofu, soybean curd from soybean milk, is consumed throughout Asia. It was eaten in the tenth century in China and Japan. *Tofu* is traditionally consumed after it is produced with a combination of magnesium dichloride ($MgCl_2$) as *Nigari* and calcium dichloride ($CaCl_2$) as *Sumashi-ko*. More recently, glucono-δ-lactone (GDL) has been added to it for commercial production. Tofu resembles cheese or yogurt made from milk curd of cows or other mammals. It is made from soy milk. It curdles by *Nigari* or a coagulant agent [15]. The forming system of *Tofu* curdling is shown in Fig. 2.

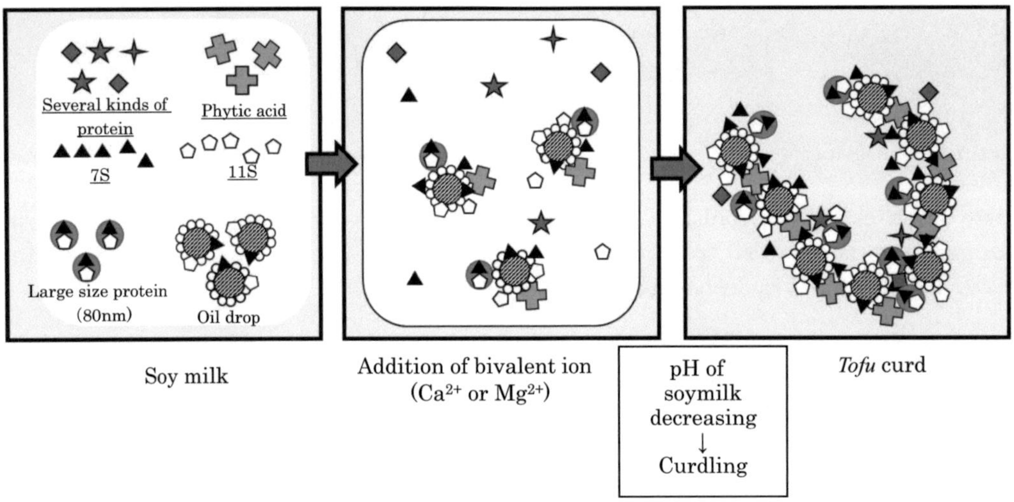

Figure 2. Formation of *Tofu* curdling [15].

Soy milk is an emulsion composing lipid and soy protein, mainly glycinin and β-conglycinin. Actually, 60% of the protein in soy milk protein is composed of these two proteins, and a large size of the protein body was constructed with their proteins and others. Lipid is a triacylglyceride composed of linoleic acid, oleic acid, and phospholipids. Before treating, triacylglyceride is stored in oil bodies in soybeans. During soy milk production, and of course during *Tofu* production, oil drops are suspended, forming an emulsion after crushing of soybeans. The oil drop is formed and held stably by lecithin, phospholipid, and oleosin, all proteins forming oil bodies in soybeans. Furthermore, an outside layer is covered by soy protein. Phytic acid (*myo*-inositol-1,2,3,4,5,6-hexaphosphate) and mineral and oligosaccharide are in the soluble component of soy milk [16]. When some bivalent ions such as calcium ion and magnesium ion are added to soy milk, the ions combine with phytic acid. As a result, decreasing the pH of the soybeans immobilizes the protein in it [17].

This phenomenon induces the charge of protein to dissipate by a combination of phytic acid and bivalent ion. Repulsion among their proteins is decreased. Moreover, the immobilized protein combines with outside layer protein easily around an oil drop [18]. Consequently, the steric network structure of soy protein is formed to gather oil drops with intermediary soy protein. Becoming low pH in soy milk, soluble soy protein becoming dissoluble is taken into the network. The curd produces a hard gel that is water retentive, such that moisture is trapped in the network.

4. Soy milk curdling yeast and characteristics of curdling enzymes

4.1. Soy milk curdling

As discussed above, soybean proteins provide rich nutrition [4, 19]. In fact, soymilk consumption is increasing quickly throughout the world because of its health benefits. Moreover, differently from bovine milk, it contains no cholesterol. Yogurt-like foods and cheese-like foods made from soybeans can be consumed by people who are concerned about health issues or allergies related to bovine milk. Tofu, soybean curd from soybean milk, resembles cheese or yogurt made from milk curd of cows or other mammals. However, the cheese mouthful sense and physical properties are not identical to those of casein protein. Unlike the protein casein in bovine milk, enzymatic curdling of soybean milk produces poor flavor and texture.

It is not yet a viable alternative to dairy foods. For that reason, the commercial use of enzymes such as bromelain, ficin, and papain for the curdling of soybean milk has been unsuccessful [20–22]. The authors have reported that physical properties of soy protein modified by enzyme reactions such as germinated proteolysis in soybeans. Therefore, in this section, along with a report of yeast containing soybean curdling enzyme [23], this investigation was undertaken to screen and identify specific food yeast strains (*Saccharomyces* sp.) that produce a soybean milk curdling enzyme and to purify the enzyme using chromatographic procedures.

4.2. Screening of yeast producing curdling soybean milk enzyme

The yeast strains (1345 strains) stored in the laboratory were screened using soybean milk agar plate medium. The strains were inoculated by streaking on a plate surface. Then they were incubated at 30°C for 7 days. After cultivation, the clear zone diameter was measured using calipers. Yeast strains that produced a clear zone were selected. Results show that 1242 yeast strains among all 1345 yeast strains produced no clear zone on the plate medium. The yeast strains (42 strains) produced less than 1 mm of a clear zone. Also, 57 yeast strains produced 1.0–5.0 mm; 4 yeast strains produced more than 6 mm.

In the second screening of curdling soybean milk enzyme-producing yeast, the screened strains (103 strains) were inoculated to the soybean milk liquid medium. Purchased soybean milk was added to them aseptically.

The soybean milk medium was incubated at 30°C for 24 h. When curdling occurred, the pH of whey was measured using a pH meter (Horiba Ltd.). Results show that three yeasts curdled at pH greater than 5.90. The media were pH 5.90 (SCY 001), pH 6.05 (SCY 002), and pH 6.38 (SCY003) (Table 4).

pH	Curdling soy milk		
	(++)	(+)	(–)
≥6.50	0	0	7
6.49–5.90	1	2	43
5.89–5.50	0	17	17
5.49–5.00	2	11	0
≤4.99	3	0	0

Initial pH was 6.70. ++, coagulating very well; +, coagulating; –, noncoagulating.

Table 4. Curdling soybean milk condition by screened yeast

The curd activity of strain SCY003 was the highest among the strains. Therefore, the SCY003 strain was finally screened. Isolated yeasts were classified taxonomically and were identified according to methods described in earlier studies [24].

The morphology was observed by microscope. Their 1.5- to 6.5-μm-long cells were short and ovaloid. The yeast, which buds by multibudding reproduction, does not form pseudomycelia or pellicles on the liquid medium. It forms ascospores. It was identified as *Saccharomyces* sp.

For researching physiological characteristics of the strain, the yeast was inoculated into a yeast nitrogen base medium (Difco Laboratories) adding 0.5% of each carbon source: sugar or organic acid as glucose, galactose, sucrose, maltose, raffinose, trehalose, lactose, melibiose, cellobiose, melezitose, starch, D-xylose, L-arabinose, D-ribose, L-rhamnose, erythritol, D-mannitol, salicin, inositol, dulcitol, ethanol, D-sorbitol, disodium succinate, and trisodium citrate. The yeast was inoculated into yeast carbon base medium (Difco Laboratories) adding sodium nitrate solution.

The glucose, galactose, sucrose, maltose, and raffinose in the medium were fermented as carbon sources using strain SCY003. The yeasts grew in a vitamin-free medium. Furthermore, the strain did not grow in 0.01% cycloheximide. According to the morphological, physiological, and molecular characteristics, it was identified as *Saccharomyces bayanus*.

For researching molecular biological characteristics of the strain, primers were used for amplification and sequencing of 18S-rRNA-encoding genes. The PCR products were sequenced using a kit (ABI Prism Big Dye Terminator Cycle Sequencing Ready Reaction; Applied Biosystems). Analyses of DNA sequence reactions were performed using a sequencer (3130; Applied Biosystems). The 18S rRNA coding DNA was sequenced. Homology was assessed using the Basic Local Alignment Search Tool (BLAST; http://www.ncbi.nlm.nih.gov/BLAST/).

As a result, the yeast showed homology of 99% with *S. bayanus* (accession no. AY046227). It was identified as *S. bayanus* through homology research and phenotypic testing.

4.3. Purification of the protease as a soybean milk curdling enzyme

Enzyme extraction, intercellular, in *S. bayanus* SCY003, soybean milk curdling test/activity was conducted using the method described [25-27] for modified soybean milk from bovine milk-curdling activity. The mixture was centrifuged at 400×*g* for 10 min. The supernatant was removed gently using a Pasteur pipette. The weight of the precipitate was measured using a chemical balance.

Generally, commercial soy milk has dispersion stability attributable to the presence of oleosomes or forming aggregate formation of soy proteins on it [28, 29]. Therefore, no precipitate is produced from commercial soybean milk by low centrifugal gravity as 400×*g*. However, precipitation ratios increase with the enzyme reaction period.

The precipitation ratio was related with the reaction period, and with the enzyme solution volume. They are mutually correlated: $R^2 = 0.9978$. Therefore, one curdling unit expressed the ratio of curdling (%) from 1 mL of soybean milk at 40°C for 1 h. The precipitation ratio from soybean milk was assayed efficiently to propose a new curdling method. The precipitation ratio was assayed (Fig. 3).

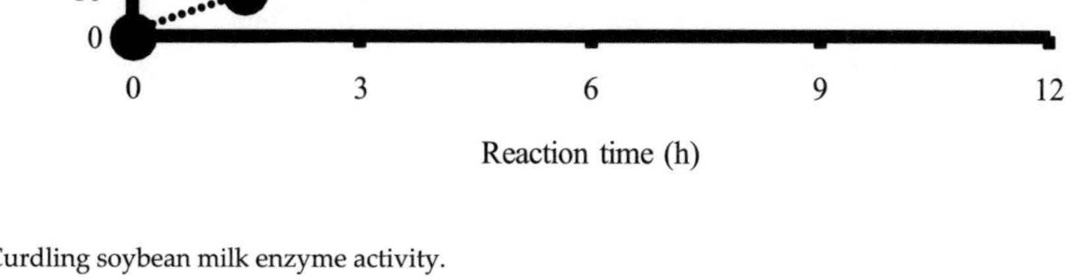

Figure 3. Curdling soybean milk enzyme activity.

After crude extraction of the enzyme, the enzyme protein was purified using chromatography. After crushing cells and extracting the enzyme, the enzyme was precipitated to between 30% and 40% saturation of ammonium sulfate. After redissolving the precipitate, the solution was dialyzed overnight at 4°C, and they carried out ion-exchange chromatography using a column (25 mm × 300 mm) of DEAE gel. The protein was eluted using a 0- to 1-M NaCl linear gradient. Proteolytic activity and curdling were assayed each fraction. Proteolytic activity was measured in duplicate using a commercial kit fluorescein isothiocyanate-labeled casein (FTC). Fluorescence was measured using a 485-nm excitation wave and a 535-nm emission wave. The proteolytic activity was decided for one unit expressed equal amount of trypsin (1 ng·mL^{-1}) doing proteolysis of FTC solution.

After ion chromatography, fractions containing the highest level of activity were pooled and reprecipitated using 80% saturation of ammonium sulfate. After redissolving the precipitate, gel filtration chromatography (10 mm × 350 mm, P-100 gel; Bio-Rad Laboratories Inc., CA, USA) was carried out. Then the molecular weight of the enzyme was analyzed using also the chromatography as various molecular weight standards (myosin, 200 kDa; serum albumin, 66.2 kDa; ovalbumin, 45.0 kDa; trypsin inhibitor, 21.5 kDa).

The results are portrayed in Figs. 4a and 4b. One main curdling activity peak was identified using ion-exchange chromatography. The peak (fraction no. 49) agreed with protein and activity. Furthermore, curdling activity agreed with the same fractions presenting protease activity (fraction no. 49).

As a result, the larger peak of proteolysis activity was found around fraction number 25, and a small peak was found at fraction number 49. The fraction of the proteolysis enzyme around number 25 was not representative of curdling activity. It is considered that the former fractions are attributable to intense proteolysis enzymes and that the latter fractions are attributable to soybean milk-curdling enzymes.

After reprecipitation, the sample was analyzed using gel filtration chromatography. A peak was found at fraction numbers 11–14. Their fractions agreed with soybean milk curdling activity, proteolytic activity, and protein. This result on their chromatograms demonstrates that protease and soybean milk curdling enzyme have some mutual relation of activity.

After purification, the enzyme protein band was approximately 45 kDa (Fig. 5a), which agrees with data of other proteases. The protease molecular weight was measured using gel filtration chromatography (Fig. 5b). The molecular mass is about 45 kDa. The protease was inferred.

The soy milk curdling enzyme has proteolytic activity. Results suggest that the soy milk curdling enzyme was a proteolysis enzyme. Many researchers have reported protease produced by yeasts as *Candida albicans* [30, 31, 32], *Candida humicola* [33], and *Saccharomyces cerevisiae* [34]. Extracellular proteases produced by yeasts as *Candida* spp. are 42–45 kDa [35, 32]; those by bacteria are 21 kDa [36].

By contrast, few reports describe intracellular protease producing *Saccharomyces cerevisiae,* although many intracellular proteases in the vacuole or other organelles are known to be related to proteinase A, which is 42 k Da [34]. The molecular weight of curdling soy protein enzyme protease agreed with protease produced by other yeast as Ascomycota. However, the *Mucor* sp. enzyme, which curdles bovine milk, produced a 49-kDa protease [37], which is larger than those produced by these yeasts.

4.4. Characteristics of the protease as a soybean milk curdling enzyme

Optimum pH, temperature, and stability of the enzyme are presented in Figs. 6a and 6b. Optimum pH was assayed at pH 4.0–8.0 using 50 mM phosphate-citric buffer or phosphate-

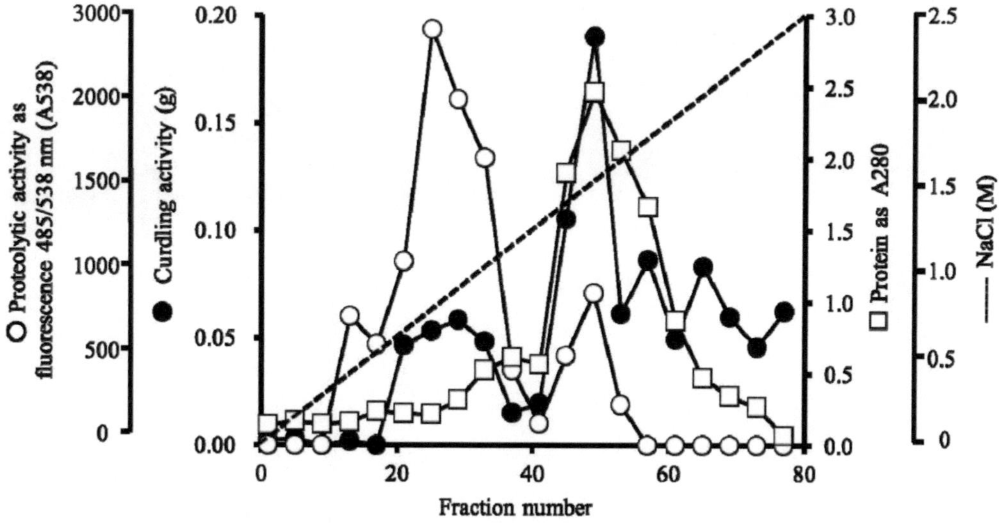

(a)

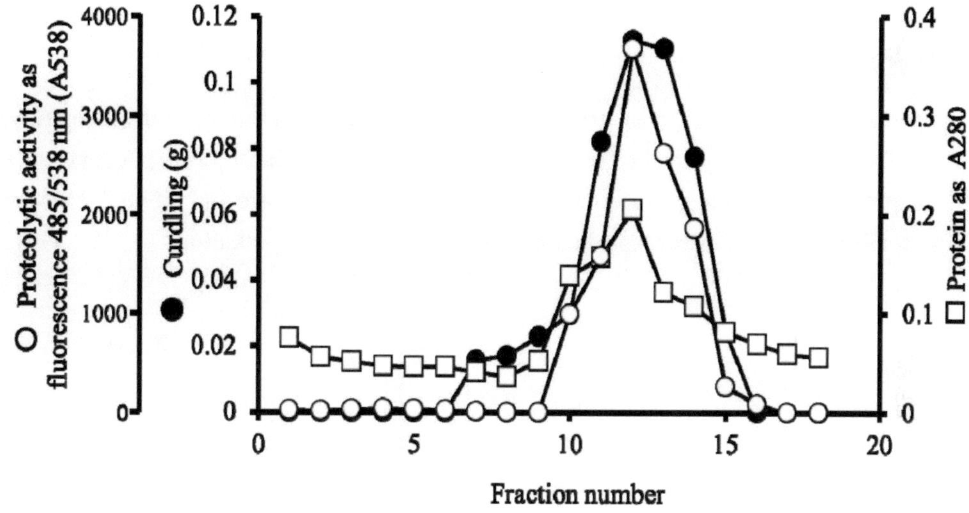

(b)

Figure 4. (a). Ion-exchange chromatography of soybean milk curdling. **enzyme.** (b). Gel filtration chromatography of soybean milk curdling. **enzyme.**

NaOH. For optimum pH of the enzyme assaying, 0.1 mL of a fluorescein isothiocyanate-labeled casein (FTC) solutions, which dissolved in each phosphate buffer (50 mM, pH 4.0–8.0), and 20 μL of enzyme solution were reacted at 40°C for 60 min. For optimum temperature of the enzyme assaying, FTC solution at pH 7.5 (0.1 mL) and 20 μL of enzyme solution were reacted at 15°C–70°C for 60 min to find optimum temperature. After reaction, fluorescence was measured using a 485-nm excitation wave and a 535-nm emission wave.

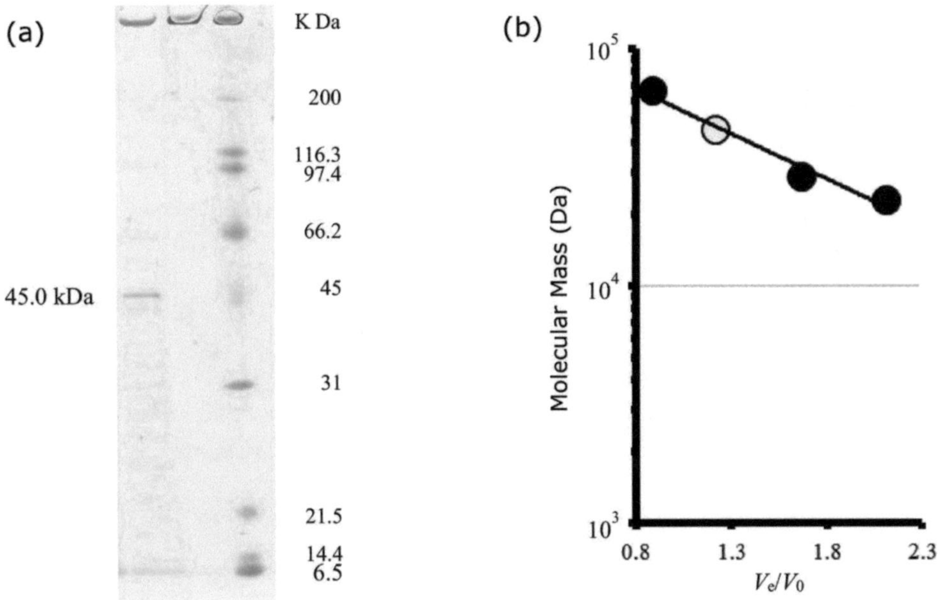

Figure 5. (a). Photograph of SDS-PAGE of soybean curding enzyme. (b). Measurement of molecular weight.

The optimum pH for the protease activity as curdling was pH 7.5; the optimum temperature was 50°C. The optimum pH of a bovine milk curdling protease, *Mucor pusillus,* is pH 5.0. The optimum pH values of many commercially available proteases are pH 5.9–6.7. However, soymilk curdling activity decreases concomitantly with increasing alkalinity.

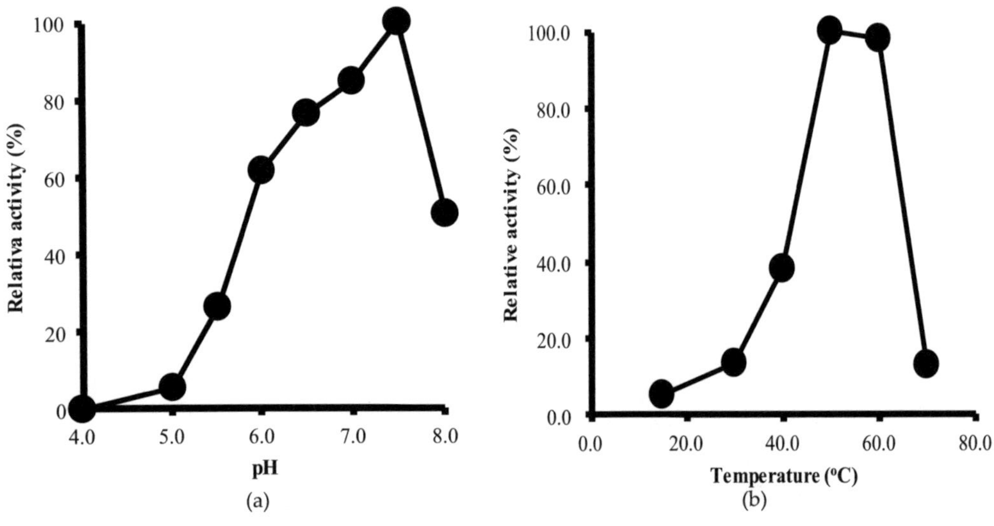

Figure 6. (a). Optimum pH of soybean curdling enzyme. (b). Optimum temperature of soybean curdling enzyme.

Park *et al.* [38] reported that the optimum pH of soybean milk curdling protease produced by *Bacillus* was pH 6.0. Regarding this enzyme, the optimum pH of the protease was pH 7.5. The optimum temperature of the protease was 50°C. The enzyme also curdled soybean milk at pH 7.5 and 50°C. Commercial soybean milks sold in Japan are pH 7.0–7.2. The pH range agrees with their optimum pH range.

Effects of metal ions and inhibitors on protease are presented in Table 5. In fact, zinc, copper, and mercury all inhibit protease activity. The amino acid of the active site contains cysteine residue because of inhibition by mercury [39]. Furthermore, EGTA (10 mM) inhibited protease activity (62.0% of relative activity).

Metal ion and inhibitor	Concentrations	Relative activity (%)
Na^+	1 mM	101.5
K^+	1 mM	105.9
Mn^{2+}	1 mM	101.9
Mg^{2+}	1 mM	46.2
Fe^{2+}	1 mM	95.4
Zn^{2+}	1 mM	23.6
Co^{2+}	1 mM	47.9
Cu^{2+}	1 mM	17.9
Ca^{2+}	1 mM	100.9
Iodo acetate	1 mM	74.1
Hg^{2+}	1 mM	34.1
EGTA	1 mM	101.6
EGTA	10 mM	62.0
EDTA	1 mM	96.1
EDTA	10 mM	89.8
NEM	1 mM	98.5
Azid	1 mM	100.4
NBS	1 mM	97.6
Cont.	–	100.0

Table 5. Effects of metal ion and inhibitor of the protease

The activity was not activated by metal ions, but it was inactivated by mercury. Soybean milk-curdling enzyme [38] was inhibited by zinc ions and mercury ions. These results agree with our data related to zinc and mercury. Its survival activity was 18% by mercury. The protease was not activated by metal ions, which indicates that the protease is not a metalloprotease: a metal-dependent enzyme. The amino acid of active site contains cysteine residue.

The mechanisms of curdling soybean milk protease were investigated. At first, The curdled soybean milk samples with added protease and without protease were treated with sample buffer solution.

Soybean milk was poured into a glass vessel (inner diameter 32 mm, height 45 mm). After 0.1 mL of enzyme solution adding to the soybean milk, or without enzyme 0.1 mL D.W., the mixtures were incubated at 40°C, and they were sampled sequentially; between 4- and 24-h. samples (0.01 mL) were added to 0.01 mL of sample buffer and then heated at 100°C for 3 min. Then samples (10 μl) were added in each well. The samples were electrophoresed on a 12.5% uniform gel at 20 mA.

They were subsequently examined using SDS–PAGE (Fig. 7) of curdled soybeans. The left side lane shows the standard of protein size. The next lane (0 h.) shows soybean milk protein without reaction of protease. The other lanes show soybean milk protein decomposed for 4, 8, 12, 16, and 24 h. Lane 0 h shows the α'- and α-subunits of β-conglycinin (approximately 84–73 kDa), the A_3 acidic subunit (approximately 40 kDa), other acidic subunits as A_4, A_{1a}, A_{1b}, and A_2 (approximately 30–42 kDa) of glycinin, the β subunit (approximately 50 kDa), and basic subunits as B_3, B_{1a}, B_{1b}, and B_4 (approximately 20 kDa) [40, 41].

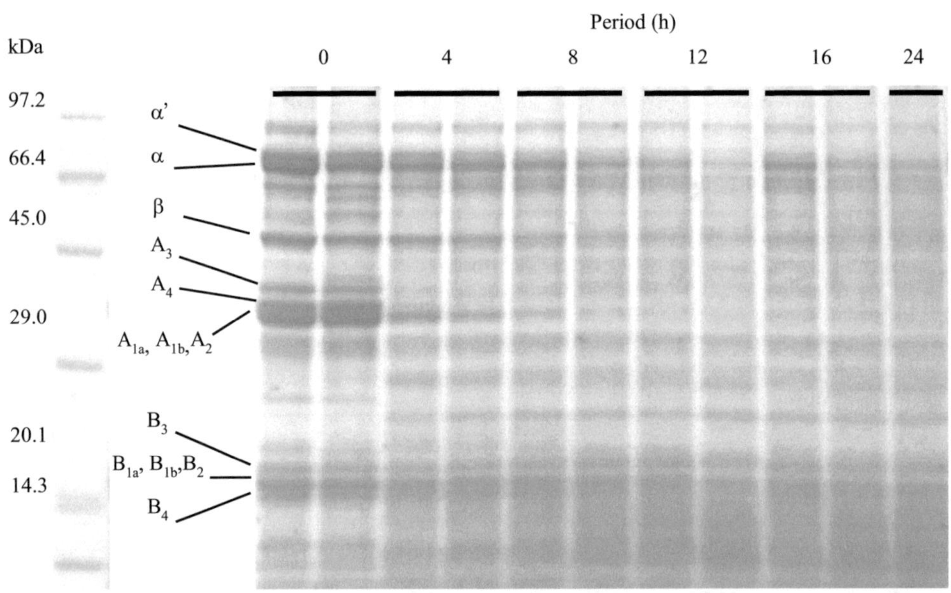

Figure 7. Digestion of soy protein during curding by soybean milk curdling enzyme.

The two bands shown as α' and α disappeared gradually after the reaction, showing the protein band from curd making the protease. In the glycinin subunits, the band of the A_3 acidic subunit disappeared completely after 4 h. Furthermore, A_4, A_{1a}, A_{1b}, and A_2 disappeared to a partial degree same as A_3 acidic subunit. Peptides smaller than 20 kDa were detected on the gel. The β-conglycinin as α, α′, and part of glycinin as A_3 A_4, A_{1b}, and A_2 were decomposed. Soybean protein became loosely curdled with the addition of other proteases from microorganisms or plants. The protein was decomposed. The low-molecular-weight peptides increased on the polyacrylamide gel. Generally, 11S glycinin was related to the formation of a stiffer gel. Furthermore, Ono *et al.* [42] reported hydrophobic bonding and hydrogen bonding related to curdling *Tofu*. Utsumi *et al.* [43] reported that the basic subunit and β-subunit formed macro complexes by heating. The complexes were regarded as forming cores for *Tofu* coagulation. The complexes were reportedly wrapped in α-, α′-, and acidic subunits [42].

According to our data, after the curdling soy milk by enzyme, α- and α′-subunits cleaved by the protease easily, whereas basic and β-subunit remained. It is considered that surface proteins as α- and α′-subunits were decomposed easily. Some decomposed subunits such as α, α′, A_3, acidic, and basic subunits are regarded as related to the curdling soybean milk.

The enzyme of mechanisms for proteolysis was searched that the enzyme had the peptidase activity as exotype proteolysis activity and protease as endotype proteolysis activity. The synthesis peptide substrates, Z-glutamyl-tyrosine, and casein, FTC, were reacted by the enzymes. Peptidase (carboxypeptidase) activity was determined by the increase in ninhydrin after hydrolysis of benzyloxycarbonyl-glutamyl-tyrosine (pH 8.0) at 40°C.

The results show that the enzyme had 0.14 $U \cdot mg^{-1}$ protein as peptidase activity and 0.55 $U \cdot mg^{-1}$ protein protease as endotype proteolysis activity (data not shown). Results also show that the soybean milk-curdling enzyme as a proteolysis enzyme had endotype proteolysis activity.

Jones *et al.* [34] reported proteolysis enzymes of three types in yeast classes: cytosolic protease, vacuolar proteases, and proteases located within the secretory pathway. They belong to aspartic type, serine, or metallo-type proteolysis enzyme cleaved substrates with endotype or exotype. Generally, metallo-type enzyme requires metal ions such as zinc. The optimum condition of aspartic protease is an acidic condition. The enzyme did not require ion metal. Its optimum pH was 7.5, which is weakly alkaline. It is therefore considered that the enzyme is a serine protease of one kind. These results agreed with serine protease from *Bacillus* sp. curdling soy protein [44]. For future studies, we will ascertain the amino acid sequence in a substrate cleaved by enzyme using synthesis substrate.

4.5. Characteristics of curd curdling by enzyme

As mechanisms that are closely involved in curdling soybeans, curdled soybean milk by enzyme and glucono-δ-lactone (GDL, 3.0% solution) as control samples were dissolved in chemical solutions. The enzyme solution (0.1 mL) was added to the soybean milk. The mixture was incubated at 40°C for 4 h., or 0.1 mL of glucono-δ-lactone (GDL, 3.0% solution) was added to the soybean milk (1.0 g) and incubated at 80°C for 1 h as control sample. To each of the two curdled soybean milk samples, 9 mL of chemical solution as 2% SDS solution, 4 M urea solution, and 10 mM 2-mercaptoethanol were added. They were held for 1 h at room temperature and were then centrifuged at 4000×*g* for 10 min. Protein in the supernatant was assayed using the Lowry method.

Results show the relative ratio of protein (%); that is, 100% relative ratio represents the amount of protein dissolving no-curdling soy milk in each solution (Fig. 8).

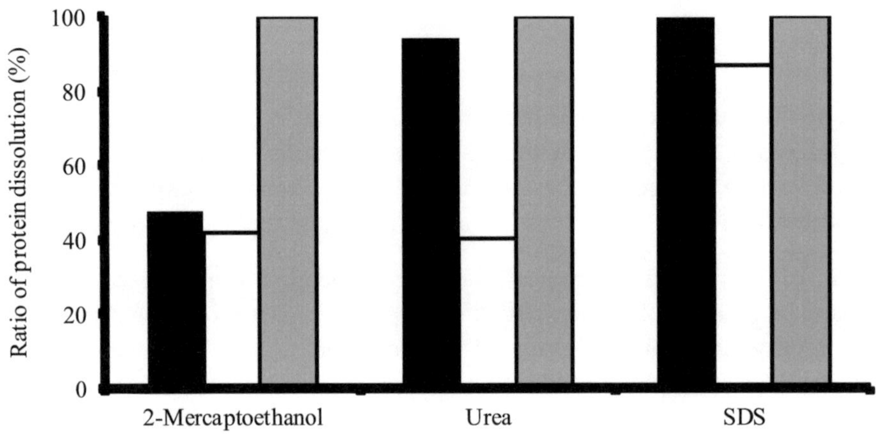

▪, SCE; □, GDL; ▪, no-curdled soy milk.

Figure 8. Curdled soy bean milk dissolved in chemical solutions. SCE shows soymilk curdled by enzymes; GDL shows soymilk curdled by glucono δ-lactone.

After dissolving the solutions, curd produced by the enzyme dissolved 47.3% of relative ratio from curd to the 2-mercaptoethanol solutions. That curdled by GDL dissolved 41.6% of relative ratio from the curd to the urea solutions. With urea solution, the curd making the enzyme dissolved 93.8% of relative ratio and GDL dissolved 40.3% of relative ratio. The curd produced by the enzyme can dissolve with urea. Both the curd produced by the enzyme and GDL dissolved with SDS solution. Actually, 2-mercaptoethanol solution cleaves the disulfide bond in protein. The urea solution cleaves the hydrogen bond, and the SDS solution cleaves the hydrophobic bond. That inference agrees with results reported by Yasuda *et al.* [44] that serine protease from *Bacillus* sp. curdled soybean milk and produced a protein bond through mutual hydrophobic bonding.

Next, the effects of the proteolysis enzyme against two protein in soybean, glycinin and β-conglycinin, were researched. Glycinin and β-conglycinin were extracted from commercial soy protein according a process described by Nagano *et al.* [44]. The soy protein (100 g) suspended 1500 mL of distilled water at pH 7.5 adjusted 0.1 M NaOH. From their extraction, the glycinin was carried out to do isoelectric precipitation at pH to 6.4. Moreover, β-conglycinin was also precipitated at pH 4.8. The two fractions were freeze-dried. Each fraction as glycinin fraction and β-conglycinin (50 mg) was resolved to 1 mL of 50 mM phosphate buffer (pH 7.5). Furthermore, enzyme solution (0.1 mL) was added. The mixture were incubated at 40°C. After reaction, soybean milk curdling activity was assayed according the preceding method. Curdling activity was assayed against two substrates: glycinin and β-conglycinin (Fig. 9). Glycinin was curdled strongly: soybean curdling activity was 86.9 ($U \cdot mL^{-1} \cdot min^{-1}$). However, β-conglycinin was curdled weakly: 38.0 ($U \cdot mL^{-1} \cdot min^{-1}$).

The data agree with reports in the relevant literature [45, 46, 47]. Bromelain decomposes 11S globulin to curdling. The entire band of acidic subunits and most basic subunits disappeared [46]. Glycinin-rich soybean milk was curdled strongly [47]. However, the enzyme made glycinin curdle without metal ion or GDL. Generally, glycinin is known to contain more sulfur amino acid than β-conglycinin does. According to Fig. 7, soymilk was curdled by a hydrogen bond or hydrophobic bond. Furthermore, some alkaline protease as subtilisin and chymotrypsin cleaves hydrophobic amino acid residue.

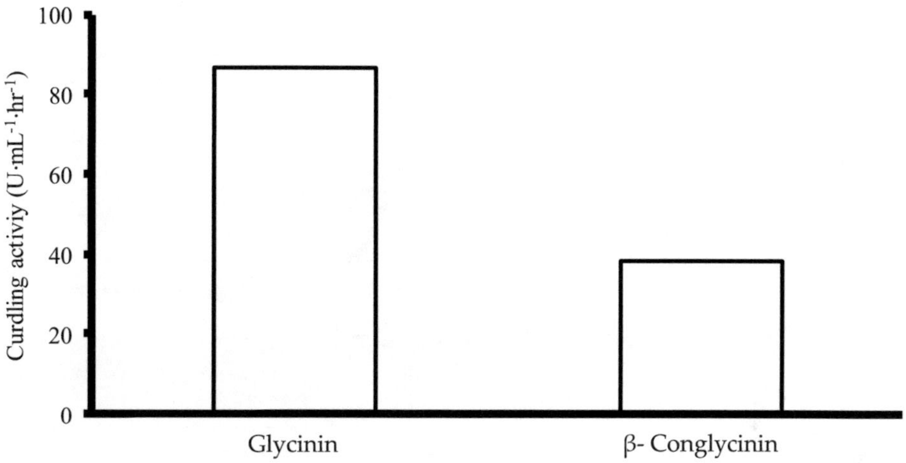

Figure 9. Curdling activity against two soy protein as glycinin and β-conglycinin.

The enzyme made curd from soymilk mainly by an enzyme reaction against glycinin. Choi *et al.* [48] reported that the α-subunit in β-conglycinin contains a hydrophobic sequence.

Soybean milk (10 mL) with 0.01 mL of anti-foam (KM-72F) was poured into a glass cup (32 mm inner diameter, 45 mm height). Then 1.5 mL of enzyme solution was added to the soybean milk. The mixture was incubated at 40°C for 4 h. As a control sample, glucono-δ-lactone (GDL, 0.3%) was added to soybean milk (10 mL) with 0.01 mL of anti-foam (KM-72F) added. Then the mixture was incubated at 80°C for 1 h. The curdled soybean milk samples were held at room temperature for 30 min. The rheological characteristics of enzyme curdling soybean milk were measured directly using a creep meter with a 16-mm-diameter plunger compressing 1 mm s^{-1} with 80.0%. As a control sample, soybean milk was curdled by GDL, 0.3% at 80°C for 1 h.

The rupture strength of the curdled soybean milk in the cups was measured directly using a creep meter (RE-3305; Yamaden Co. Ltd., Tokyo, Japan) with a 16-mm-diameter plunger compressing 1 mm s^{-1} with 80.0%.

The stress–strain curves of curdled soybean milk are presented in Fig. 10. The vertical axis shows stress ($N \cdot m^{-2}$), which represents internal forces of the sample curd pushing back against the strain. The horizontal axis shows the strain of the curd. The sample curd strained by the plunger is broken by a force that exceeds a certain force: the breaking point. The breaking load represents the hardness or softness of the curd sample. The breaking strain represents the resilience of the sample curd: a large value signifies a high-resilience sample. Pressure by the plunger was loaded more. Then the curd was broken more heavily. After strain loading, the stress value decreased partly. The brittleness shows a different breaking point with the local minimal value. A large brittleness load shows brittle sample curd. The soybean milk was poured into a glass vessel and then curdled using the respective methods. After curdling, rheological analysis of the sample was conducted using a creep meter without taking out.

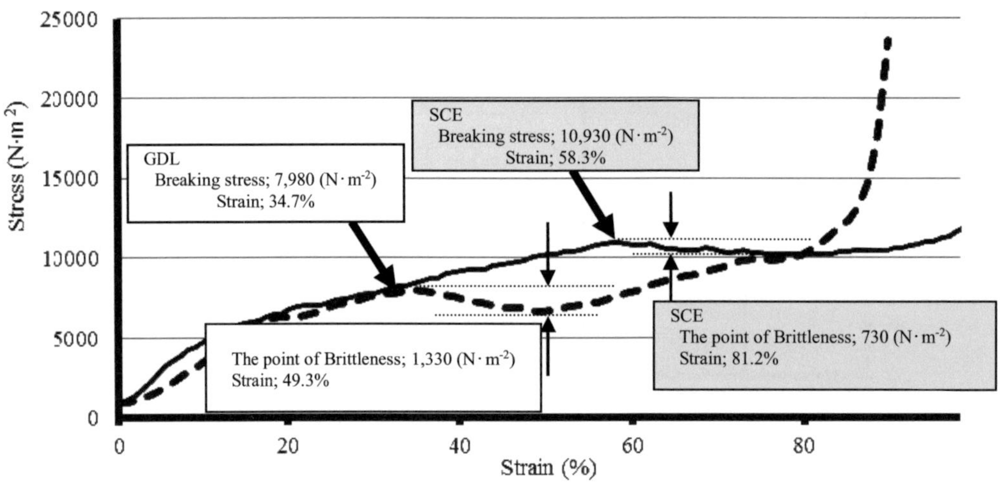

Figure 10. Stress–strain curves of curdled soybean milk.

The breaking point of curd curdled using GDL, which is used by many Japanese *Tofu* production companies to produce commercial *Tofu*, was 34.7% strain. The breaking stress was 7980 ($N \cdot m^{-2}$). The point of brittleness was 49.3% and 6650 ($N \cdot m^{-2}$). The brittleness of this curd produced using GDL was 1330 ($N \cdot m^{-2}$). In fact, their *Tofu* curdled by GDL same as the Japanese *Tofu* is soft and brittle.

By contrast, the breaking point of curd curdled by soy curdling enzymes (SCE) was 58.4% strain. Their breaking stress was 10,930 ($N \cdot m^{-2}$). The brittleness point is 81.2% and 10,200 ($N \cdot m^{-2}$). The brittleness of this curd produced using the enzyme was 730 ($N \cdot m^{-2}$). The curd that used SCE had 1.4 times greater breaking point, shown hardness, than that of the curd produced using the GDL same as the Japanese *Tofu*. Moreover, the curd that used SCE had 1/2 times smaller brittleness than that of GDL. Great breaking point shows hard curd and elasticity. It showed that the curd produced by SCE had hard, but soft, springy and sticky curd.

Heretofore, Yasuda *et al.* [45] reported soy milk curdled by bacteria protease. As a result, the curd produced using the bacteria protease is too soft to measure rheological characteristics, but they also reported that curd produced using the bacteria protease with calcium ion had 2–3×10^5 ($N \cdot m^{-2}$) of the breaking stress. Their curd was as hard as 20–30 times of the curd produced using this yeast enzyme. The curd produced using the bacteria protease with calcium was broken during fermentation and aging for long time. However, the curd produced using the bacteria protease had not springy and sticky texture.

It is considered that the curd produced using yeast enzymes was not like *Tofu* or the curd produced using the bacteria protease of its main characteristics or rheology.

Guo and Ono [47] and Toyokawa *et al.* [49] reported that the breaking stress of normal *Tofu* showed a relationship to soy milk conditions, such as their concentrations of glycinin, proteins, or temperature. Future investigations will examine if the condition of soy milk has a relationship with the breaking stress for the enzyme curdling.

According to this report, the new protease from *S. bayanus* SCY 003 produced a new texture of soy food that is applicable for new healthy foods, anti-milk allergy foods, and others.

5. Conclusion

Soybeans are a traditional food in eastern Asia, particularly in Japan and China. They were eaten in 100 BC in China. The beans can be processed into *Tofu*, soy milk, fermented seasonings, soy sauce or *Miso* paste, and *Natto* and green beans. Soybeans have rich nutrition, protein lipid, and other functional substances such as isoflavones. However, soybeans are difficult to process for use as food because of tissue and cell wall hardness. Therefore, soybeans are conducted to do some treatments, *e.g.*, boiling, steaming, roasting, crushing/grinding, and some enzyme treating, to eat soy protein easily.

Soy storage proteins mainly comprise two proteins as 7S globulin composed of β-conglycinin and γ-conglycinin and 11S globulin containing glycinin composed of 5 subunits. β-Conglycinin, included in 7S globulin, is composed of three subunits.

To modify the physical properties of soy protein, a new type of enzyme for curdling soybean milk enzyme was purified as an extract from yeast. Yeast producing curdling soybean milk enzyme, the SCY003 strain, was isolated from 1345 yeast strains. According to the morphology, physiology, and molecular and characteristics, SCY003 was identified as *S. bayanus*. The soy milk curdling enzyme having proteolytic activity was approximately 45 kDa and a monomer protein. The optimum pH for the protease activity was pH 7.5; the optimum temperature was 50°C.

The enzyme cleaved the β-conglycinin as α and α', and part of glycinin as A_3 A_4, A_{1b}, and A_2 in soy protein by endoproteolysis. Soybean protein became loosely curdled with the addition of other proteases from microorganisms or plants. Soybean milk curdled after cleaving endoproteolysis enzyme in SCY003 strain. The rheological characteristics of enzyme curdling soybean milk, the breaking point, was 58.4% strain; their breaking stress was 10,900 ($N \cdot m^{-2}$); the brittleness point is 81.2% and 10,200 ($N \cdot m^{-2}$). The brittleness of the curd produced using the enzyme was 727 ($N \cdot m^{-2}$). The curd had a sticky and chewy texture and did not resemble *Tofu* rheologically.

In this way, some properties of soy protein were modified by enzymes, such as decomposing specific subunits in soybeans and making soy milk curdle, which are expected to be applicable for new healthy foods, anti-milk allergy foods, and others.

Acknowledgements

We thank A-STEP, Adaptable and Seamless Technology, Transfer Program through Target Driven R&D by Japan Science and Technology Agency (JST) (AS231Z00291E) for financial support.

Author details

Makoto Kanauchi[1], Sakiko Hatanaka[2] and Makoto Shimoyamada[3]

1 Department of Food Management, Miyagi University, Japan

2 Industrial Technology Institute, Miyagi Prefectural Government, Japan

3 School of Food and Nutritional Sciences, University of Shizuoka, Japan

References

[1] Yamauchi F, Okubo K. editors. Daizu no Kagaku: Science of Soy Bean. Asakura Publishing, Tokyo: 1992 (in Japanese).

[2] The yield amount of soy bean; http://www.maff.go.jp/j/seisan/ryutu/daizu/d_data/pdf/014.pdf

[3] Trend in demand of soy bean; http://www.maff.go.jp/j/seisan/ryutu/daizu/d_data/pdf/011.pdf

[4] Endres JG. In: Soy Protein Products: Characteristics, Nutritional Aspects, and Utilization. Champaign, Illinois: AOCS Publishing; pp.10–18 (2001).

[5] Food and Drug Administration (FDA) in U.S. Soy Protein and Coronary Heart Disease. In: Federal Register. 1999:57700–57733.

[6] CMC Publishing. β-Conglycinin. Bio industry, 27 (5): 68–69 (2010) (in Japanese).

[7] Yamori Y, Ohta S, Watanabe S. Soybean Isoflavones. Tokyo: Saiwai Shobo; 2001 (in Japanese).

[8] Fritz H, Seely D, Flower G, Skidmore B, Fernandes R, Vadeboncoeur S, Kennedy D, Cooley K, Wong R, Sagar S, Sabri E, Fergusson D. Soy, red clover, and isoflavones and breast cancer: a systematic review. PLoS One 2013;8(11): e81968.

[9] Simons R, Vincken JP, Roidos N, Bovee TF, van Iersel M, Verbruggen MA, Gruppen H. Increasing soy isoflavonoid content and diversity by simultaneous malting and challenging by a fungus to modulate estrogenicity. J Agric Food Chem 2011;59(12): 6748–58.

[10] Zimmermann MC, Tilghman SL, Boué SM, Salvo VA, Elliott S, Williams KY, Skripnikova EV, Ashe H, Payton-Stewart F, Vanhoy-Rhodes L, Fonseca JP, Corbitt C, Collins-Burow BM, Howell MH, Lacey M, Shih BY, Carter-Wientjes C, Cleveland TE, McLachlan JA, Wiese TE, Beckman BS, Burow ME. Glyceollin I, a novel antiestrogenic phytoalexin isolated from activated soy. J Pharmacol Exp Ther 2010;332(1): 35–45.

[11] Kraft B, Barz W. Degradation of the isoflavone biochanin A and its glucoside conjugates by *Ascochyta rabiei*. Appl Environ Microbiol 1985;50(1): 45–8.

[12] Tikhonovich IA. Nitrogen Fixation: Fundamentals and Applications: Fundamentals and Applications. Proceedings of the 10th International Congress on Nitrogen Fixation, St. Petersburg, Russia; 1995.

[13] Takashima M, Nara K, Niki E, Yoshida Y, Hagihara Y, Stowe M, Horie M. Evaluation of biological activities of a groundnut (*Apios americana* Medik) extract containing a novel isoflavone. Food Chem 2013;138(1): 298–305.

[14] Lin LZ, He XG, Lindenmaier M, Yang J, Cleary M, Qiu SX, Cordell GA. LC-ESI-MS study of the flavonoid glycoside malonates of red clover (*Trifolium pratense*). J Agric Food Chem 2000;2(48): 354–65. doi:10.1021/jf991002

[15] Shimoyamada M. Processing characteristics of soy protein—curdling property of soy milk. J Cookery Sci Japan 2007;40(1): 37–40 (in Japanese).

[16] Ono T, Takeda M, Guo ST. Interaction of protein particles with lipids in soybean milk. Biosci Biotechnol Biochem 1996;60: 1165–9.

[17] Ono T, Katho S, Mochizuki K. Influences of calcium and pH on protein solubility in soybean milk. Biosci Biotechnol Biochem 1993;57: 24–8.

[18] Guo ST, Ono T, Mikami M. Incorporation of soy milk lipid into protein coagulum by addition of calcium chloride. J Agric Food Chem 1999;47: 901–5.

[19] Larkin T, Price WE, Astheimer L. The key importance of soy isoflavone bioavailability to understanding health benefits. Crit Rev Food Sci Nutr 2008;48(69): 538–52.

[20] Park YW, Kusakabe I, Kobayashi H, Murakami K. Production and properties of a soymilk-clotting enzyme system from a microorganism. Agric Biol Chem 1985;49(11): 3215–9.

[21] Park YW, Kobayashi H, Kusakabe I, Yoshida S, Murakami K. Action of soymilk-clotting enzyme from *Bacillus* sp. K-295G-7 on the acidic subunit of soybean 11S globulin. Agric Biol Chem 1989;53(8): 2289–90.

[22] Qua DV, Shimizu U, Taga N. Purification and some properties of halophilic protease produced using a moderately halophilic marine *Pseudomonas* sp. Can J Microbiol 1981;27: 505–10.

[23] Hatanaka S, Maegawa M, Kanauchi M, Kasahara S, Shimoyamada M, Ishida M. Characteristics and purification of soybean milk curdling enzyme-producing yeast *Saccharomyces bayanus* SCY003. Food Sci Technol Res 2014;20(5): 927–38.

[24] Kurtzman CP, Fell JW. In: The Yeast, A Taxonomic Study. 4th ed.. Amsterdam, The Netherlands: Elsevier Science Publishers B.V., pp.360–361 and 891–913, 1998.

[25] Arima K, Yu J, Iwasaki S, Tamura G. Milk-clotting enzyme from microorganisms. V. Purification and crystallization of Mucor rennin from *Mucor pusillus* var. Lindt. J Appl Microbiol 1968;16: 1727–33.

[26] Khan MR, Blain JA, Patterson, JDE. Extracellular protease of *Mucor pusillus*. Appl Environ Microbiol 1979;37(4): 719–24.

[27] Nouani A, Belhamiche N, Slamani R, Belbraout S, Fazouane F, Bellal MM. Extracellular protease from *Mucor pusillus*: purification and characterization. Int J Dairy Technol 2009;6(1): 112–7.

[28] Waschatko G, Junghans A, Vilgis TA. Soy milk oleosome behaviour at the air–water interface. Faraday Discuss 2012;158: 157–69.

[29] Shimoyamada M, Tsushima N, Tsuzuki K, Asao H, Yamauchi R. Effect of heat treatment on dispersion stability of soymilk and heat denaturation of soymilk protein. Food Sci Technol Res 2008;14: 32–8.

[30] Remold H, Fasold H, Staib F. Purification and characterization of a proteolytic enzyme from *Candida albicans*. Biochim Biophys Acta 1968; 167: 399–406.

[31] Ruchel R. Properties of a purified proteinase from the yeast *Candida albicans*. Biochim Biophys Acta 1968;659: 99–113.

[32] Negi M, Tsuboi R, Matsui T, Ogawa H. Isolation and characterization of proteinase from *Candida albicans*: substrate specificity. J Invest Dermatol 1984;83: 32–6.

[33] Ray MK, Uma DK, Seshu KG. Shivajo, S. Extracellular protease from the Antarctic yeast *Candida humicola*. Appl Environ Microbiol 1992;88(6): 1918–23.

[34] Jones EW. Three proteolytic systems in the yeast *Saccharomyces cerevisiae*. J Biol Chem 1991;266(13): 7963–6.

[35] Ruchel R. Properties of a purified proteinase from the yeast *Candida albicans*. Biochim Biophys Acta 1968;659: 99–113.

[36] Cowan DA, Daniel RM. Purification and some properties of an extracellular protease (caldolysin) from an extreme thermophile. Biochim Biophys Acta 1982;705: 293–305.

[37] Nouani A, Belhamiche N, Slamani R, Belbraout S, Fazouane F, Bellal MM. Extracellular protease from *Mucor pusillus*: purification and characterization. Int J Dairy Technol 2009;6(1): 112–7.

[38] Park YW, Kobayashi H, Kusakabe I, Murakami K. Purification and characterization of soymilk-clotting enzymes from *Bacillus* sp. K-295G-7. Agric Biol Chem 1987;51(9): 2343–9.

[39] Springham DG, Moses V, Cape RE. In: Biotechnology, the Science and the Business. New York: Harwood Academic Publishers; 1999.

[40] Thanh VH, Shibasaki K. Heterogeneity of b-conglycinin from soybean seeds. Biochim Biophys Acta 1976a;439: 326–38.

[41] Thanh VH, Shibasaki K. Major proteins of soybean seeds: Subunit structure of b-conglycinin. J Agric Food Chem 1978;26: 692–5.

[42] Ono T, Wada T, Imai A. The structure of Tofu for preventing the change of lipid. Daizu Tanpakushitsu Kenkyu 2004;7: 42–7 (in Japanese).

[43] Utsumi S, Damodaran S, Kinsella JE. Heat-induced interactions between soybean proteins: preferential association of 11S basic subunits and β subunits of 7S. J Agric Food Chem 1984; 32: 1406–12.

[44] Nagano T, Hirotsuka M, Mori H, Kohyama K, Nishinari K. Dynamic viscoelastic study of the gelation of 7S globulin from soybeans. J Agric Food Chem 1992;40: 941–4.

[45] Yasuda M, Kuba M, Tachibana S, Aoyama M. Analysis for mechanism of soybean-milk-coagulation by bacterial protease and utilization of the enzyme to the food processing. Daizu Tanpakushitsu Kenkyu 2002;5: 36–40 (in Japanese)

[46] Mohri M, Matsushita S. Improvement of water absorption of soybean protein by treatment with bromelain. J Agric Food Chem 1984;32(3): 486–90.

[47] Guo S, Ono T. The role of composition and content of protein particles in soymilk on Tofu curding by glucono-δ-lactone or calcium sulfate. Food Chem Toxicol 2005;70(4): 258–62.

[48] Choi SK, Adachi M, Utsumi S. Improved bile acid-binding ability of soybean glycinin A1a polypeptide by the introduction of a bile acid binding peptide (VAWWMY). Biosci Biotechnol Biochem 2005;68: 1980–3.

[49] Toyokawa T, Uehara M, Mochizuki T, Tamamura T, Higa K. Comparisons of physiological and chemical characteristics with Okinawa Tofu and Japanese Tofu. Report of Okinawa Industrial Technology Center. Okinawa, Japan: Okinawa Industrial Technology Center; 2008;11: 7–11 (in Japanese).

Section 2

Food Preservation

Chapter 4

The Role of Yeast and Lactic Acid Bacteria in the Production of Fermented Beverages in South America

Fábio Faria-Oliveira, Raphael H.S. Diniz,
Fernanda Godoy-Santos, Fernanda B. Piló,
Hygor Mezadri, Ieso M. Castro and
Rogelio L. Brandão

Additional information is available at the end of the chapter

http://dx.doi.org/10.5772/60877

Abstract

Fermentation is one of the oldest forms of food preservation in the world. In South America, most fermented beverages are nondairy products featuring several other food raw materials such as cereals, fruits, and vegetables. Generally, natural fermentations are carried out by yeast and lactic acid bacteria forming a complex microbiota that acts in cooperation. Yeast have a prominent role in the production of beverages, due to the ability to accumulate high levels of ethanol and to produce highly desirable aroma compounds, but lactic acid bacteria are particularly important in fermentation because they produce desirable acids, flavor compounds, and peptides that inhibit the growth of undesirable organisms. Among the South America beverages based on cereals and vegetables, the fermented beverages *chicha, caxiri, cauim* and *champús,* and *cachaça,* a fermented and distilled beverage, could be cited. Genetic and physiological analyses of *Saccharomyces cerevisiae* strains isolated from *cachaça* have been shown to present interesting traits for beer production, such as flocculation and production of aroma compounds, fundamental to high-quality beer. The study of these traditional beverages allows the identification of new microorganism strains displaying enhanced resistance or new flavor and aroma profiles that could lead to applications in several industries and ultimately new products.

Keywords: yeast, lactic acid bacteria, fermented beverages, South America, *cachaça*

1. Introduction

Alcoholic beverages have been consumed by mankind since ancient times. These products of fermented sugar-rich goods, namely, cereals, roots, and fruits, are present worldwide since the oldest records [1, 2]. In fact, several of mankind's milestones, such as the dawn of agriculture, are closely linked with the production of some type of alcoholic beverages. Similar processes of fermentation emerged independently in many civilizations across the globe. Interestingly, the main players of the whole process are relatively few, mostly yeast from the *Saccharomyces* genus and lactic acid bacteria (LAB) [3, 4]. Nowadays, such microorganisms have a significant role in several industrial relevant processes, including the production of beer, wine, cheese, and bread. Importantly, the popularity of fermented beverages, namely, beer and wine, is such that their worldwide consumption is second only to nonalcoholic drinks as water, tea, and coffee [5].

This chapter aims to contribute to a comprehensible analysis of the role of yeast and LAB on the production of fermented beverages from South America. The microbiological diversity associated with the fermentation of a wide diversity of raw materials, from sugarcane to cassava, as well as new potential biotechnological applications will be addressed.

1.1. Ethanol and lactic acid fermentation

1.1.1. Yeast diversity and metabolism

Yeast are unicellular fungi, being the simplest eukaryotes. Present in a great number of environments, yeast can be found not only in decomposing fruit, trees, and soils but also in commensal relationships with higher eukaryotes, humans included, and even saltwater. The high diversity of species, almost 1500 species have been described [6], is closely related to this wide distribution. Some of these yeast are adapted to extreme environments, such as high salt concentrations [7], low pH [8], or extremely cold temperatures [9, 10]. The genus *Saccharomyces*, particularly *Saccharomyces cerevisiae*, is strongly associated with the production of fermented products for human consumption, namely, bread, wine, and beer [2]. After several millennia of close coexistence, through phenotypic selection, these species evolved to produce goods with organoleptic properties pleasant to humans. However, given the high degree of diversity found in nature, it is expected to find yeast with new and more interesting characteristics for the industry in new and unexplored niches [11, 12].

Yeast, as other heterotrophic organisms, have the anabolism coupled with catabolism. In one hand, the oxidation of organic molecules, as sugars, yields adenosine 5-triphosphate (ATP) that, in turn, is used as an energy resource for the cell. On the other hand, such organic molecules can also be used as building blocks or to generate intermediary compounds for the synthesis of other compounds, some of which with high commercial value.

The high diversity of environments where yeast can be found is closely related to the variety of carbon sources that can be used. Hexoses such as glucose, fructose, galactose, or mannose are the most common substrates, but some species can use pentoses like xylose or arabinose.

http://dx.doi.org/10.5772/60877

Several industrial relevant species can metabolize disaccharides as maltose, lactose, or sucrose, and some, as *Saccharomyces diastaticus*, can even metabolize dextrins (glucose polymers) [13, 14]. Nevertheless, glucose and fructose, to a lesser extent, are the preferred substrates.

In order to use glucose as carbon source, first and foremost, yeast have to sense the presence of this sugar in the surrounding environment and then express the adequate proteins to transport it across the plasma membrane [15, 16]. Whenever glucose is sensed in the medium, changes in the cell proteome will occur. Several processes contribute to the overall change in enzymes levels, including alteration of mRNA translation rates, mRNA stability, or protein synthesis and/or degradation. However, the major response is the extensive upregulation of a large number of genes required for the metabolism of glucose, such as genes encoding glycolytic pathway enzymes, leading to the adaptation to the fermentative metabolism. Moreover, in genes encoding for proteins involved in the metabolism of alternative substrates, gluconeogenic and respiratory pathways are repressed strongly by glucose (for reviews, see [17, 18]). In *S. cerevisiae*, a glucose concentration as little as 15 mM is enough to induce such changes [19].

S. cerevisiae presents an extensive family of hexose transporters, including more than 20 members: (i) 18 genes encoding transporters (*HXT*1-*HXT*17, *GAL*2) and (ii) at least two genes encoding sensors (*SNF*3, *RGT*2). Some studies suggest that Gpr1p and Hxk2p may sense glucose levels [17, 20]. The transporters can be divided in two classes regarding glucose affinity: (1) low affinity for glucose and high transport capacity, the most important proteins are Hxt1p and Hxt3p, and (2) high affinity and low transport capacity, the key proteins being Hxt2p, Hxt4p, and Hxt7p.

Following uptake by the hexose transporters, glucose enters the glycolytic pathway in order to be metabolized to pyruvate (Figure 1, steps from glucose to pyruvate), whereby the production of energy in the form of ATP is coupled to the generation of intermediates and reducing power in the form of NADH for biosynthetic pathways [21, 22]. The phosphorylation of glucose to glucose-6-phosphate, requiring ATP, is the initial step of glycolysis, by the action of the hexokinases (Hxk1/2p) and the glucokinase (Glk1p), which are linked to high-affinity glucose uptake. The glucose-6-phosphate is then isomerized to fructose-6-phosphate by the phosphoglucose isomerase, encoded by *PGI1* gene. The next step, done by the phosphofructokinase (Pfk1/2p), also requires energy. The fructose-6-phosphate molecule is converted into fructose 1,6-biphosphate through the transfer of inorganic phosphate from ATP. In turn, yeast aldolase (fructose 1,6-bisphosphate aldolase—Fba1p) is responsible for the reversible cleavage of fructose 1,6-bisphosphate to glyceraldehyde 3-phosphate and dihydroxyacetone phosphate.

These two resulting compounds can be interconverted, in a reversible way, by the action of the triosephosphate isomerase (Tpi1p). Glyceraldehyde 3-phosphate is further metabolized to ultimately yield pyruvate, while some of the dihydroxyacetone phosphate follows gluconeogenesis. This step is fundamental for the osmotic and redox homoeostasis, as the dihydroxyacetone can be converted to glycerol yielding NAD^+. Glyceraldehyde 3-phosphate is first oxidized by NAD^+ and then phosphorylated under the catalysis of the 3-phosphate dehydrogenase (Tdh1/2/3p). The resulting 1,3-diphosphoglycerate is, in turn, converted to 3-phosphoglycerate by the action of phosphoglycerate kinase (Pgk1p), yielding 1 molecule of ATP. The enzyme phosphoglycerate mutase (Pgm1p) promotes the relocation of the phosphate group from C3 to C2, allowing the dehydration by the enolase (Eno1/2 p), resulting in the phosphoenolpyruvate. Then the pyruvate kinase (Pyk1p) converts this highly energetic molecule to pyruvate, yielding a second molecule of ATP.

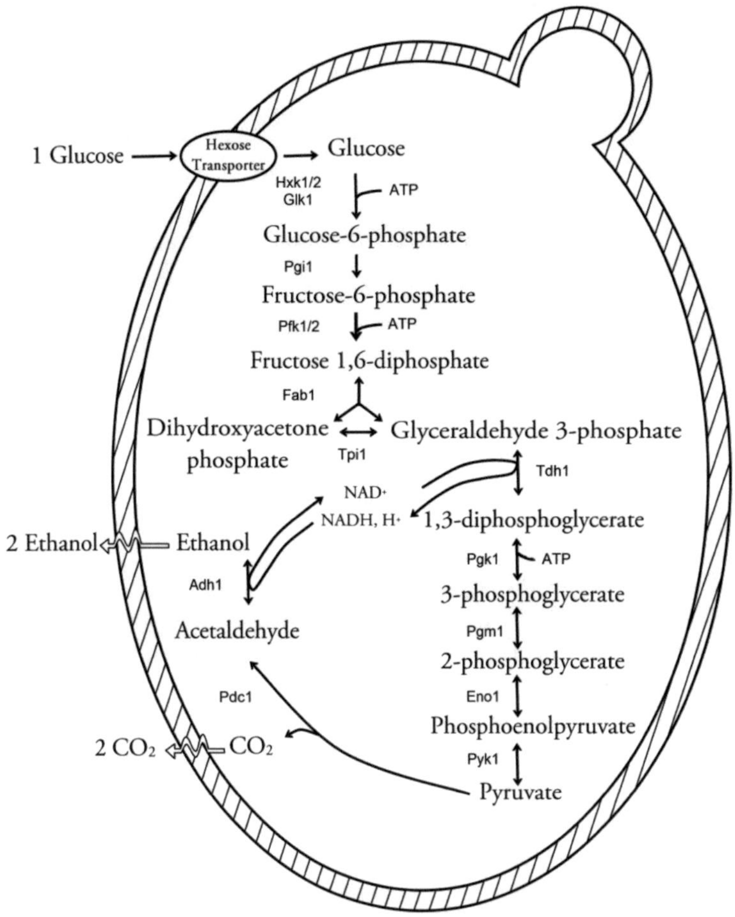

Figure 1. Glycolysis and alcoholic fermentation steps on *S. cerevisiae* (adapted from [23]).

The pyruvate molecule can be further processed through different metabolic alternatives, the respiratory or the fermentative pathways (Figure 2). The selection of one of the route depends greatly on the expression or repression of some genes, which in turn are tightly regulated on the environmental conditions [24]. The genus to which the yeast belongs also plays a role in the prevalence of one route over the other.

The fermentative pathway is particularly relevant to industry, as several important commodities are produced through this process (characteristic of particular organisms). In *S. cerevi-*

siae, the first step is the decarboxylation of pyruvate to yield acetaldehyde and carbon dioxide (CO_2), through the action of the pyruvate decarboxylase (Pdc1/5/6p). The acetaldehyde can be further reduced to form ethanol by the enzyme alcohol dehydrogenase (Adh1p), allowing the reoxidation of NADH to NAD^+. Besides the direct products of fermentation, ethanol and CO_2, several other by-products are generated during the process, including cell biomass, glycerol, and some organic acids. Overall, the ethanol fermentation is a redox-neutral process since the reduced coenzyme NADH produced during glycolysis, in the oxidation of glyceraldehyde 3-phosphate, is latter reoxidized in the reduction of acetaldehyde to ethanol [25]. Nevertheless, given that biomass is a product of fermentation, and it is in a more oxidized state than glucose, an excess of reducing equivalents may be generated. As mentioned above, glycerol production plays an important role in the redox balance restoration. The glycolytic intermediate dihydroxyacetone is reduced to glycerol 3-phosphate, oxidizing NADH to NAD^+, in a reaction catalyzed by the NAD^+-dependent glycerol 3-phosphate dehydrogenase (Gpd1/2p). Glycerol 3-phosphate is then dephosphorylated to glycerol due to the action of glycerol 3-phosphatase (Gpp1/2p) [5, 26, 27]. The presence of glycerol may contribute to the organoleptic properties in the final product of fermentation, such as wine.

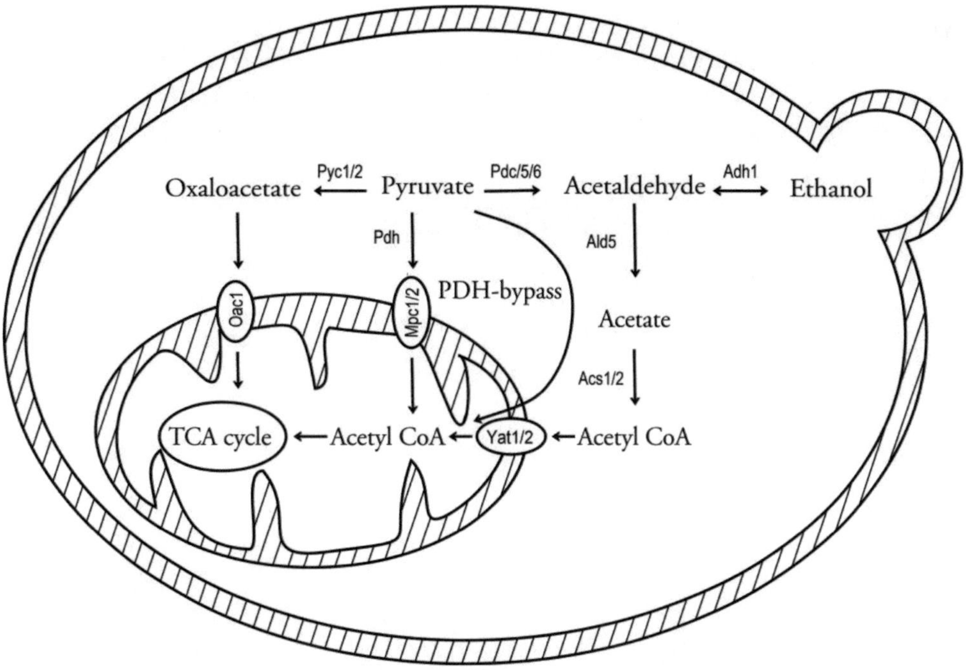

Figure 2. Pyruvate metabolic fates. The pyruvate yielded during glycolysis can be converted into two intermediates of TCA cycle: acetyl-CoA, by the pyruvate dehydrogenase complex (PDH), and/or oxaloacetate, by pyruvate carboxylases (Pyc1/2p). These molecules can be transported into the mitochondria by the pyruvate carriers (Mpc1p or Mpc2p) and the oxaloacetate carrier (Oac1p), respectively. Another alternative is the decarboxylation to acetaldehyde, by the pyruvate decarboxylase (Pdc1/5/6p), which ultimately can yield ethanol. Adh1p—alcohol dehydrogenase; Ald5p—acetaldehyde dehydrogenase; Acs1/2p—acetyl-CoA synthases; Yat1/2p—carnitine acetyltransferases (adapted from [22]).

Although most microorganisms ferment in the absence of oxygen, this is not always the case. Even if oxygen is available, high concentrations of sugars present in the environment will lead yeast to choose fermentation over respiration. This inhibition of aerobic metabolism if glucose is available, both in the presence or absence of oxygen, is denominated the Crabtree effect [28]. *S. cerevisiae* is known as Crabtree positive since it will produce ethanol aerobically if the glucose available is higher than 15 mM [19]. The availability of high sugar concentrations in the surrounding environment stimulates glycolysis, which in turn leads to the production of increasing amounts of ATP, through substrate-level phosphorylation. At the same time, the availability of additional ATP will reduce the respiration and ATP synthesis, through oxidative phosphorylation, leading to a decrease in oxygen consumption. On the other hand, Crabtree-negative yeast do not present a glucose inhibition of aerobic respiration, so these microorganisms resort to this more efficient form of energy metabolism, producing biomass via tricarboxylic acid (TCA) cycle. Nevertheless, these species are able to ferment, but mainly in anaerobic conditions. Importantly, Crabtree is not exclusive to yeast, as it has been detected in many mammalian tumor cells [29–31].

During aerobic respiration (Figure 3), acetyl-CoA is produced by the decarboxylation of the glycolytic pyruvate, by the action of the pyruvate dehydrogenase complex. Then acetyl-CoA will enter the tricarboxylic acid (TCA) cycle, where it will be used to generate reducing equivalents, NADH and $FADH_2$. These molecules will fuel the oxidative phosphorylation, through the highly conserved electron transport chain. Besides the production of reducing coenzymes, the TCA cycle provides intermediates to several other biochemical pathways, including the synthesis of amino acids and nucleotides (for reviews, see [22, 32]).

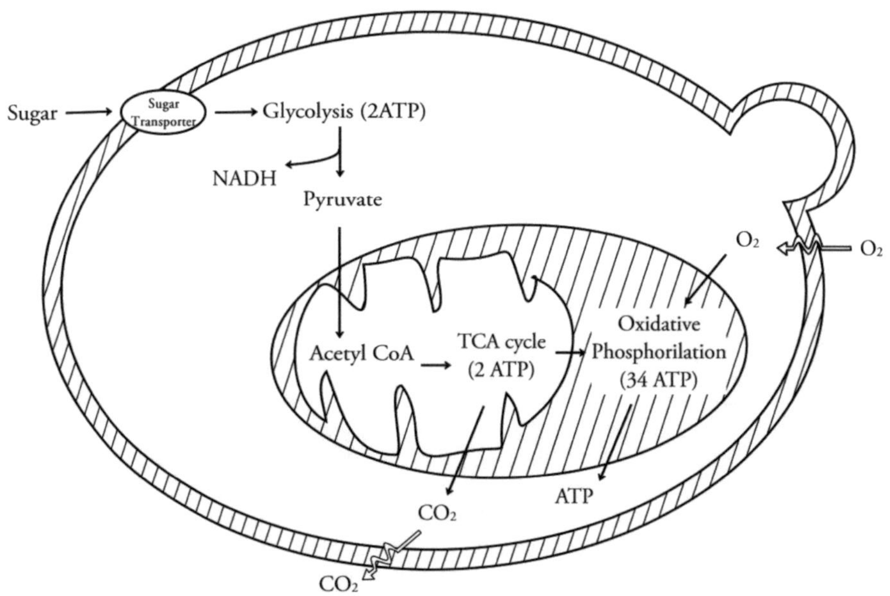

Figure 3. Aerobic respiration in *S. cerevisiae* (adapted from [33]).

1.1.2. Lactic acid bacteria

Lactic acid bacteria (LAB) constitute an ubiquitous and heterogeneous group capable of fermenting carbohydrate with the production of lactic acid as a major end product [34]. LAB are found in diverse nutrient-rich habitats associated with plant and animal's matter, as well as in respiratory, gastrointestinal, and genital tracts of humans [35, 36]. A typical LAB is Gram positive, present a GC content below 55%, generally nonsporulating, usually nonmotile, fastidious, catalase negative (pseudocatalase may occur in some LAB), aerotolerant, and acid tolerant [34]. Taxonomic parameters have distributed LAB members into two phyla, *Firmicutes* and *Actinobacteria*. Within the *Firmicutes* phylum, LAB members belong to the order *Lactobacillales* and comprise the following genera: *Aerococcus, Alloiococcus, Carnobacterium, Enterococcus, Lactobacillus, Lactococcus, Leuconostoc, Oenococcus, Pediococcus, Streptococcus, Symbiobacterium, Tetragenococcus, Vagococcus*, and *Weissella*. Regarding LAB members belonging to the *Actinobacteria* phylum, the only species belongs to the *Bifidobacterium* genus [35, 37, 38]. Nevertheless, it is worth mentioning that *Bifidobacterium* is poorly phylogenetically related to typical LAB. These bacteria have been considered as LAB given its physiological similarity and the shared biochemical properties [39, 40].

Usually, LAB members are nonpathogenic organisms with a reputed generally recognized as safe (GRAS) status. The *Lactobacillus* genus includes some of the most important GRAS species involved in food microbiology and human nutrition [41, 42]. The remarkable ability of these bacteria to adapt to different environments resulted in a large number of industrially relevant strains. Among these are *Aerococcus, Carnobacterium, Enterococcus, Lactobacillus, Lactococcus, Leuconostoc, Pediococcus, Streptococcus*, and *Bifidobacterium* [35, 43, 44]. Furthermore, given that LAB greatly contribute to the effective acidification of the matrix and consume rapidly fermentable sugars, these bacteria are frequently predominant in the natural fermentation microbiota [44].

1.1.2.1. Pathway of homolactic and heterolactic acid fermentation in LAB

LAB are able to live in the presence of oxygen; however, they obtain their energy by substrate-level phosphorylation. These bacteria do not present a functional respiratory system, as they lack the ability to synthesize cytochromes and porphyrins, key components of respiratory chains [45, 46]. Therefore, an important parameter used in the differentiation of the LAB species is the type of lactate fermentations: homofermentative and heterofermentative [35]. As a general rule, homofermentative lactic acid bacteria use the Embden–Meyerhof–Parnas pathway (EMP pathway or glycolysis) to produce pyruvate, while heterofermentative lactic acid bacteria use the pentose phosphate pathway (PPP). However, a third pathway, the Bifidum pathway, presents distinct reactions (Figure 4) [45, 46].

In the homofermentative lactate fermentation, as the name implies, the major end product generated is lactate. Initially, two ATP molecules are produced per mole of glucose via the oxidation of phosphoglyceraldehyde. In a second stage, NADH molecules resulting from the previous oxidative stage are used to reduce the pyruvate, forming lactate [45, 46]. The overall reaction is as follows:

$$\text{glucose} + 2\,\text{ADP} + 2\,\text{Pi} \rightarrow 2\,\text{lactate} + 2\,\text{ATP}$$

Some representative homolactic LAB genera include *Lactobacillus, Lactococcus, Enterococcus, Streptococcus*, and *Pediococcus* species [38].

Conversely, in the heterofermentative lactate fermentation pathway, lactate is not the only end product; significant amounts of CO_2 and ethanol, or acetate, are also produced. In this pathway, lactate is produced by the decarboxylation and isomerization reactions of the PPP. Glucose is oxidized to ribulose-5-phosphate that is isomerized to xylulose-5-phosphate, which in turn is cleaved to form phosphoglyceraldehyde and acetyl phosphate. The phosphoglyceraldehyde molecule is oxidized to pyruvate by reactions of glycolytic pathway, whereas the acetyl phosphate is reduced to ethanol [45, 46]. The overall reaction is as follows:

$$\text{glucose} + \text{ADP} + \text{Pi} \rightarrow \text{ethanol} + \text{lactate} + CO_2 + \text{ATP}$$

Some representative heterolactic LAB genera include *Leuconostoc, Oenococcus*, and *Weissella* [38]. It is worth mentioning that heterofermentative lactate fermentation produces only one ATP molecule per glucose, while the homofermentative lactate fermentation produces two ATP molecules per glucose.

1.1.2.2. Bifidum pathway

The Bifidum pathway is a particular metabolic route found in *Bifidobacterium bifidum*, which uses reactions of the PPP and homofermentative pathway, producing primarily acetate and lactate [38, 45]. In this pathway, 2.5 ATP molecules are produced per glucose. As such, ATP yields are greater than for the homofermentative or heterofermentative pathways, due to the presence of key enzymes, fructose-6-phosphate phosphoketolase and xylulose-5-phosphate phosphoketolase. These proteins catalyze two important steps: the cleavage of one molecule of fructose-6-phosphate, yielding one molecule of erythrose-4-phosphate and one of molecule acetyl phosphate, and the cleavage of two xylulose-5-phosphate into two glyceraldehyde 3-phosphate and two acetyl phosphate, respectively [45, 46]. The overall reaction is as follows:

$$2\,\text{glucose} + 5\,\text{ADP} + 5\,\text{Pi} \rightarrow 3\,\text{acetate} + 2\,\text{lactate} + 5\,\text{ATP}$$

1.1.2.3. LAB—beverage industry applications

Over the years, LAB has been explored on a large scale in several food industry segments (processing of meats, vegetables, and beverages) occupying a central role in these niches [43, 48–50]. Withal, there are some reasons that explain their use in the food production industry. Among these are the following: the production of antimicrobial substances, which restricts the growth of harmful microorganisms, and the production of

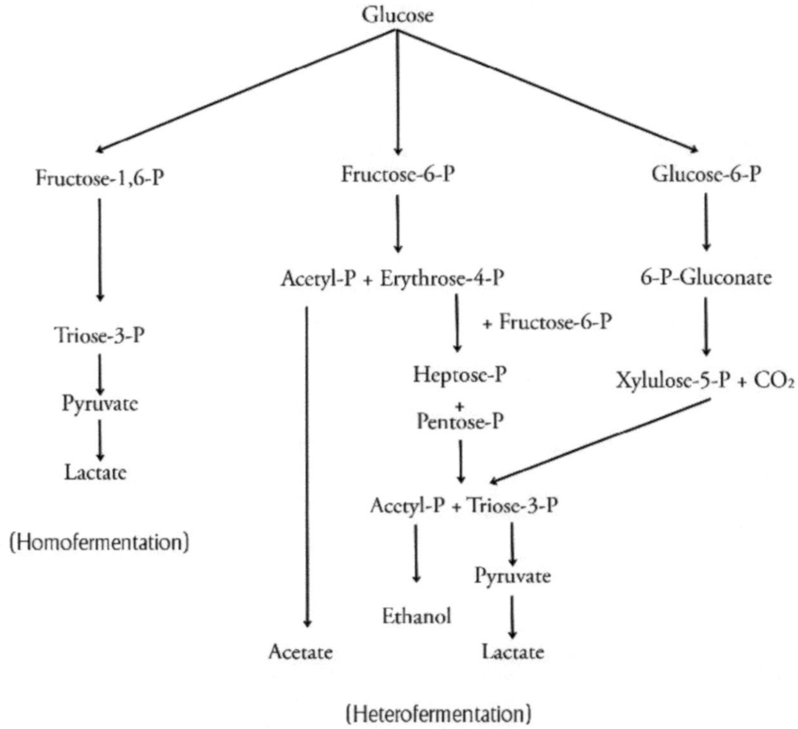

Figure 4. Schematic representation of the metabolism of hexoses by lactic acid bacteria (adapted from [47]).

metabolites, which influences the nutritional, texture, and organoleptic qualities of the end products [36, 51]. Moreover, LAB have also been used as probiotics, which shows several potential health benefit [52]. Thus, in general, LAB enhances the shelf life and microbial safety of end products [43]. However, based on the microorganisms profile present in the raw material, their effects may be either beneficial or disadvantageous to the food processing. For instances, malolactic fermentation (MLF) is a secondary fermentation in wine normally carried out by LAB, especially by *Leuconostoc oenos*, which usually occurs at the end of alcoholic fermentation by yeast [53]. In this metabolism, L-malic acid is decarboxylated to L-lactic acid and CO_2, a reaction catalyzed by the malolactic enzyme without the release of intermediates. As a consequence of this pathway, the acidity is reduced which turn it a crucial process in wine and cider production [53]. However, it is noteworthy that MLF is not only important as a deacidification process in wine, but for the aroma and microbial stability of wine [53–55]. Additionally, this fermentation prevents the malic acid utilization by other undesirable microorganisms. Another industrial application of LAB is the use of starter cultures as inoculants during the malting process, a complex biological process essential to the production of fermented beverages, in order to improve the malt quality and safety. In these conditions, LAB can improve the extraction, fermentability, and nitrogen yield of wort and the foam stability, color, and flavor of beer [50]. Moreover, another important effect of the LAB in the malting process is their ability of antimicrobial substances production (e.g., bacteriocins) that restricts the growth of harmful bacteria to malting [50, 56–58].

2. Fermented beverages of South America

The traditional foods, mainly those produced by spontaneous fermentation, are present in the daily life of the population and play an important role in the cultural identity of different communities [59]. Indigenous or traditional fermented foods refer to the products that, since the beginning of history, are an integral part of the diet and can be prepared in household or cottage industry, using simple techniques and equipments [60].

In South America, there are various traditional fermented beverages, mainly produced by fermentation of cereals, vegetables, and root tubers. Among these beverages could be cited the traditional beverages *cauim* and *caxiri*, made by Brazilian natives and the traditional beers *chicha* and *champús*, typical of the Andes [59, 61–64]. *Cachaça* is also a typical beverage, from Brazil, where the fermented sugarcane juice is then distilled to produce a spirit. Table 1 shows some characteristics of these traditional beverages.

Beverage	**Raw material**	**Microorganisms**	**Country**
cauim	cassava, rice, peanuts	Lactic acid bacteria (LAB), *Saccharomyces cerevisiae*, other yeast	Brazil
caxiri	cassava	LAB, *Bacillus* spp., *S. cerevisiae* (predominant yeast), other yeast	Brazil
champús	maize	*S.cerevisiae*, other yeast	Colombia
chicha	maize	LAB, *S.cerevisiae* (predominant yeast), *Aspergillus* spp.	Peru
cachaça	sugarcane	*S. cerevisiae;* LAB, other yeast	Brazil

Source: [59, 61-64]

Table 1. Catalogue of traditional fermented beverages of South America.

2.1. *Caxiri*

Caxiri is an alcoholic fermented beverage made from cassava (*Manihot esculenta* Crantz) and produced by Brazilian natives. The *Yudjá-Pakaya* native tribe is located in the Xingu Park in the Mato Grosso state, Brazil [64]. Initially, cassava roots are incubated for 2 days in running water to soft the skin. After this stage, the cassava is peeled, cut into small pieces, and placed in handmade straw press for removing the excess of water. The resulting paste from pressing is grated into flour. Then the flour is roasted for about 2 hours. Subsequently, the flour is mixed with water and grated sweet potato. The mixture is placed in special open vessels for fermentation to occur. After 24–48 hours, the beverage is ready for consumption [64]. In the course of the fermentation process, a change of pH occurs, decreasing from pH ~5.8 to ~3.1–3.2. The final beverage has an ethanol concentration of approximately 80 g/L, and 28 g/L of lactic acid.

Ramos et al. (2010) investigated the microbiota involved in *caxiri* by culture-dependent and culture-independent methods of cultivation. These authors found that the bacterial populations varied from 3.05 to 5.33 log CFU/mL, and the populations of yeast from 3.27 to 7.34 log CFU/mL. Yeast dominate fermentation after 48 hours, being *S. cerevisiae* the dominant yeast species. Other yeast species, such as *Rhodotorula mucilaginosa, Pichia membranifaciens, Pichia guilliermondii* and *Cryptococcus luteolus* were also found. The bacteria belonging to the genus *Bacillus* were the most prevalent in *caxiri*. LAB species were found at the beginning of the fermentation and after 24 hours of fermentation. The microbiota found in *caxiri* may be from different sources, such as raw materials, environmental, or utensils used in the preparation of the beverage [64].

2.2. *Cauim*

Cauim (Kawi) is a nonalcoholic fermented beverage produced from various substrates, such as banana, cassava, cotton seed, maize, pumpkin, peanuts, and rice. This drink is produced in Brazil, in the Mato Grosso state, by *Tapirapé-Tapi'itãwa* native tribe [59, 63]. *Cauim* represents great importance to the tribe since it is consumed by adults and children as part of their daily meal [59, 63]. For the production of *cauim* from cassava, initially this substrate is fermented for 3 to 5 days in running water. Afterward, the cassava is peeled, cut into pieces and placed on sun to dry. The dried pieces are crushed, and the obtained flour is added to water. The mixture is boiled for 2 hours and then cooled at room temperature. An inoculum obtained from the chewed sweet potato (*Ipomoea batatas*) prepared by indigenous women is then added to the porridge to start the fermentation. After 24–48 hours, the beverage is ready for consumption [59].

For the preparation of beverages with other substrates, the procedure is similar to *cauim* made from cassava. The main substrate is cooked for approximately 2 hours and cooled at room temperature. After cooled, the inoculum is added to start the fermentation [63]. During fermentation, a progressive acidification occurs, and the pH of beverage decrease from ~5.5 to ~3.4 [59].

Almeida et al. (2006) found LAB as the dominant microorganisms during *cauim* fermentation. Among the LAB obtained in that work, *Lactobacillus pentosus* and *Lactobacillus plantarum* were the most prevalent species. Gram-positive sporulating bacteria, as the genus *Bacillus* spp., were found in small amounts. In a recent work, Ramos et al. (2010) monitored the community of yeast and LAB in *cauim* prepared from rice and peanuts, by culture-dependent and culture-independent methods. LAB were found in higher counts, ranging from 7.4 to 8.4 log CFU/mL, while yeast were found at 4.0 to 6.6 log CFU/mL. The most prevalent species of yeast found were *Pichia guilliermondii, Kluyveromyces lactis, Candida* sp., *Rhodosporidium toruloides*, and *S. cerevisiae*. The most prevalent species of LAB belonged to genus *Lactobacillus*: as *L. plantarum, L. fermentum, L. paracasei*, and *L. brevis* [63].

2.3. *Champús*

Champús is a cereal-based fermented beverage, sweet and sour, and with low alcohol content. This drink, typically found in Colombia, can also be found in other countries in South America, such as Ecuador and Peru [62]. *Champús* can be prepared from different cereals (maize, rye, and wheat) alone or in a mixture, with other raw materials, such as pineapple, *lulo* (*Solanum quitoense* Lam.), *panela* (brown sugar paste), herbs (orange leaves), and spices (cloves and cinnamon) [62].

In Colombia, the beverage is produced by boiling the kernels of maize, for about 2 hours. Thereafter, the beans are cooled to room temperature, and then fruits, *panela,* and other ingredients are added. The beverage is cooled to 12°C–15°C, and after 24–48 hours of a spontaneous fermentation process, it is ready for consumption. The final beverage has a low alcohol content (2.5%–4.2%), and the pH is between 3.5 and 4.0 but can vary according to the ingredients [62]. The microorganisms responsible for *champús* fermentation (during storage at low temperatures), such as yeast and LAB, come from the fruit since the microbial derived from corn grains are eliminated during the period of boiling [62]. Osorio-Cadavid et al. (2008) found seven genera of yeast when twenty samples of *champús* from Colombia were analyzed. The most prevalent species of yeast founded in this study were *Pichia fermentans, S. cerevisiae,* and *Issatchenkia orientalis* (*Candida krusei*).

2.4. *Chicha*

Chicha is an alcoholic beverage, clear, yellowish, and sparkling, which resembles the taste of cider and that has been consumed by Andean indigenous population for hundreds of years. This beverage is produced in regions of the Andes and sometimes in low-lying regions of Ecuador, Brazil, Peru, Bolivia, Colombia, and Argentina [65]. *Chicha* is a generic name that comprises a series of fermented or nonfermented beverages that can be prepared from various raw materials such as cereals and fruits [66].

Chicha can be produced in different ways. Although the recipes pass from generation to generation, all of them use the conversion of starch into sugar, followed by fermentation of sweet wort. As the production process resembles the brewing process, the traditional *chicha,* made from maize, can be named as the Andean indigenous beer [67].

In the Andean region, the most common maize *chicha* is the *chicha de jora*. This *chicha* is prepared from yellow corn grain (*maíz amarillo*) malted (germinated and dried) or chewed. In *chicha de jora* production process, hydrolysis of starch is obtained by the malting process of the kernels of maize or the action of salivary amylase in the case of chewing [65].

The production process of *chicha de jora* is laborious. After malting, it proceeds to the wort boiling (consisting of maize flour plus water), a process that takes many hours. During the wort boiling, sugar and *panela* as well as herbs and spices may be added. Subsequently, the wort is cooled and filtrated. Then the beverage is placed in special vessels for the occurrence of fermentation. The *chicha* is ready for consumption when its sweet taste disappears and the flavor becomes a little stronger. However, if not consumed immediately, the beverage becomes bitter, and after 7 days, it usually is converted into vinegar [65, 68]. During the *chicha de jora* production, a wort acidification occurs, with a decreasing pH values of ~5.7–5.3 to ~3.7–3.5. The final beverage presents an ethanol concentration between 9 and 10 g/L [69].

In some Andean countries is produced the *chicha morada*, a beverage prepared with purple maize. Purple corn is a pigmented variety of *Zea mays* L., originating mainly from Peru and Bolivia. This drink is prepared by boiling purple maize with pineapple, quince peel, cinnamon, and cloves [70, 71].

In Ecuador, in addition to *chicha de jora*, other kind of *chichas* like *chicha de morocho*, prepared with white maize, *chicha de Yamor* or seven-grain *chicha*, and *chicha de yuca* (cassava *chicha*) are also produced. The seven-grain *chicha* is produced from seven varieties of maize as *jora*, yellow corn, white corn, black corn, *chulpi* corn, *morocho* corn, and popcorn [68]. The *chicha de yuca*, produced by the indigenous and mestizo population in the Amazon region of Ecuador, is produced in a peculiar way since chewing is used. After the chewing process and fermentation of cassava, a mixture of Ungurahua palm (*Oenocarpus bataua* subsp. *bataua*, Arecaceae) fruit juice with the fermented mass is made, and thus the beverage is ready for consumption. For the preparation of the juice, the fruits are first harvested and then they are soaked in hot water for the removal of mesocarp. The seeds are dropped and then pieces fermented cassava are added to the mixture of mesocarp and water [68, 72].

In *chichas* from Ecuador and Brazil, LAB, yeast, and *Bacillus* species were found as the microorganisms associated with this beverages [73, 74]. Blandino et al. (2003) found yeast, bacteria, and filamentous fungi in *chichas* of Peru. Elizaquível et al. (2015) found the *Lactobacillus* genus as the most prevalent and the one with the highest diversity of species in *chichas* of Argentina, by culture-dependent and culture-independent methods. In another work, the yeast obtained from *chichas de jora* collected in 10 *chicherías* (*chicha* producers) in Peru were identified as belonging to the species *S. cerevisiae* [66]. Vallejo et al. (2013) considered *S. cerevisiae* as the responsible yeast for the fermentation of Peruvian *chichas de jora*.

The *chicha* microbiota may come from different environments. The LAB found in *chichas* may have been introduced from the raw materials, as many species are commonly found in vegetables and plants, and also transferred from humans and animals, natural hosts of these bacteria [73]. The yeast involved in *chichas* production process come from different sources, such as handlers, raw materials, utensils, and equipment used in the preparation of these beverages or can be carried by insects. The clay vessels and wooden spoons, used in the preparation of *chichas*, provide an ideal microhabitat for yeast, infiltrating into tiny cavities of such utensils [75]. In Quito, Ecuador, two isolates of yeast from old vessels obtained from deep tombs of La Florida archaeological site were recovered, which were identified as *Candida theae*, a new species belonging to the clade *Lodderomyces* [76].

2.5. *Cachaça*

In Brazil, *cachaça* (ka.ʃa.sə) was the name given to (i) waste of sugar production (beginning of XVI century), (ii) waste of sugar production when fermented (around XVI–XVII centuries), and finally (iii) product of the distillation of the fermented sugarcane (XVII century to nowadays). The first mention of *cachaça* occurs in 1622 with the name of "augoa ardente" [aqua vitae] or spirit, in Bahia State (Brazil), and the first use of the name *cachaça* instead of aqua vitae occurs in 1660. Considering the three ethnic groups that formed the Brazilian nation (Native Brazilians, Africans, and Europeans), scarce information is available about the real contribution of each group to the initial production of fermented sugarcane. However, *cachaça*, which is the result of the distillation of fermented sugarcane juice, was certainly "discovered" by Europeans, the most technologically advanced group, who had knowledge and equipment to do so [77]. Nowadays, *cachaça* is the typical and exclusive denomination to Brazilian spirit produced from sugarcane juice with alcohol content ranging 38%–48% (v/v) at 20°C (68°F), which present unique characteristics (Table 2) [78]. Brazil has an estimated installed capacity of *cachaça* production ranging from 1.2 to 1.5 billion litters/year; however, the production is less than 800 million litters/year. According to the Brazilian Institute of Geography and Statistics (IBGE), almost 15,000 establishments are currently producing *cachaça*. In 2014, 10.2 million liters of *cachaça* were exported to 66 countries, generating US$ 18.33 million in revenue [79].

Compounds	Units	Limits
Copper	mg/L	5.0
Ethyl carbamate	µg/L	210.0
Volatile acidity	mg/100mL anhydrous ethanol	150.0
Total esters	mg/100mL anhydrous ethanol	200.0
Aldehydes	mg/100mL anhydrous ethanol	30.0
Total higher alcohols*	mg/100mL anhydrous ethanol	360.0
Furfural+HMF+	mg/100mL anhydrous ethanol	5.0
Methanol	mg/100mL anhydrous ethanol	20.0
Acrolein	mg/100mL anhydrous ethanol	5.0
Particles in suspension	-	Absent
Dry extract	g/L	6.0
Total sugars	g/L	6.0
		38-48

Source: Ministry of Agriculture, Livestock and Supply - Brazil

*Sum of isobutyl (2-methyl-1-propanol), isoamyl (2-methyl-1-butanol and 3-methyl-1-butanol), and n-propyl (1-propanol) alcohols.

+ 5-(Hydroxymethyl)furfural

Table 2. Components present in *cachaça* and its limits in accordance with the Brazilian law.

2.5.1. Production

The main raw material for the production of *cachaça* is the juice of sugarcane (*Saccharum* spp.). The first step to prepare of the fermentation medium is the extraction of sugarcane juice. Small producers extract the juice by crushing the sugarcane using a mill. Yet, the large producers use a more complex system of extraction: (i) crushing system, (ii) cutting machines, and (iii) shredders [80]. The resulting sugarcane juice is an opaque (color ranging from brown to dark green), viscous, and sweet liquid. The color is due to different pigments such as chlorophyll and polyphenols, while the viscosity is due mainly to the presence of colloidal proteins. The sugarcane content in fermentable carbohydrates is sucrose (11%–18%), glucose (0.2%–1%), and fructose (0%–0.6%) [81]. A great number of microorganisms is associated with the sugarcane plant, and during the extraction of juice, these microorganisms can be transferred to the fermentation medium. The yeast present in the juice belong to the genus: *Candida, Cryptococcus, Kluyveromyces, Hansenula, Rhodotorula, Saccharomyces*, and *Torulopsis*. The main bacteria genus are *Leuconostoc, Streptotococcus, Lactobacillus*, and *Bacillus* [82]. In order to reduce the number of microorganisms and/or optimize the fermentation, several processes to improve the quality of the fermentation medium can be implemented. The most common procedure is the decantation of the sugarcane juice, thus eliminating coarse particles as soil/sand that can damage the *cachaça* manufacturing equipment. A dilution or concentration step (most unusual process) may be performed, particularly if the crop was harvested at a nonoptimal time. The juice supplementation with nutrients can provide a more robust fermentation; however, such practice is not common among the majority of *cachaça* producers. Studies have shown that addition of ammonium sulfate, magnesium sulfate, cobalt sulfate, and vitamins (especially B complex) provide huge productivity gains. Other actions such as filtration, addition of antibiotics, use of disinfectants, temperature control, and must acidity correction can also be used to improve the quality of fermentations [83].

2.5.2. Microorganisms

As important as the preparation of the medium is the preparation of microorganisms that will ferment the sugarcane juice, the so-called foot-of-vat. Traditionally, *S. cerevisiae* is the most used yeast, but some other species are also utilized, namely, *Pichia* sp. [11, 84]. The cell concentration usually found at the beginning is about 10^6–10^7 cells/mL and about 10^8 cells/mL at the end of the fermentation. Traditionally, the majority of the *cachaça* is produced through self-inoculation, using different protocols. Usually, a homemade mixture is prepared: (i) diluted sugarcane juice (carbon source), (ii) rice bran and/or corn bran (nutrient sources), and (iii) lemon/orange juice (to reduce initial pH) [85]. All these ingredients are mixed and kept at rest for 24 hours, during which wild yeast present in sugarcane juice will multiply—microorganism multiplication is verified through foaming. At the end of the first 24 hours, a new volume of diluted sugarcane juice is added, and the process is restarted [85]. The process is repeated until it reaches 20% of the volume of the fermentation vat. At the end of the process, a wild yeast community is obtained, which is adapted to the physical and chemical conditions of the fermentation. Producers, which aim to improved organoleptic characteristics, tend to use selected yeast in fermentative process. There are two basic types of selected yeast: (i) yeast that are used to produce ethanol or other beverages without being specific for *cachaça* production and (ii) yeast selected from their own vats of *cachaça* fermentation. Yeast selected in their own fermentation tanks have advantages over the first because they are more adapted to the *cachaça* fermentation specific conditions and are able to provide unique features to the product [86, 87]. There are numerous advantages to using selected yeast, including (i) rapid and (ii) homogeneity of the fermentation, (iii) higher fermentation yields, (iv) higher quality of the final product, (v) lowest risk of contaminations, and (vi) highest resistance to stress [88]. The improved aroma obtained in the final product is another advantage of using the selected yeast [86]. Given that *cachaça* fermentation develops in an environment with few controls, these vats represent a unique ecological niche. Interesting phenotypes, such as the production of flavor and aroma compounds, can be assessed by the detailed characterization of isolated strains [89]. The use of yeast strains producing aroma compounds, as esters (ethyl acetate, isoamyl acetate, ethyl hexanoate, ethyl octanoate, and ethyl decanoate) and higher alcohols (propyl, isoamyl, and isobutyl alcohol), can significantly increase the final quality of *cachaça*. Thus, such profile can be a differential characteristic between several *cachaça* producers or even distinct producing regions—if all producers in a particular region start using a selected yeast strain.

2.5.3. Fermentation

The *cachaça* fermentation can occur in simple batch system (most common process) and fed batch or continuous culture (less common). The batch system with cell recycling at the end of each fermentation is the most widespread in the *cachaça* production. In general, the fermentation occurs in 24/36 hours depending on the system efficiency [90]. After fermentation, the product is generally removed by pump systems, or via a valve near the base of the fermenter, where about 80% of fermented must is removed. The remainder medium, approximately 20%, is composed primarily of sedimented yeast [90]. Some producers can perform specific treatments to decrease the contamination load of the inoculum: (i) acid treatment (Melle–Boinot method and its modifications); (ii) by stirring and spray air with diluted sugarcane juice, or (iii) only add new medium for fermentation and restart a new fermentation cycle, without any special treatment [91]. *Cachaça* production occurs mainly on small farms where financial resources are generally scarce. Control of the fermentation conditions is rarely held, including (i) maintaining the fermentation temperature, (ii) standardization of sugarcane juice, or (iii) fermentation in closed vat, in order to prevent contamination. Moreover, hygiene conditions are not always in conformity to national or international standards for the production of beverages [92]. Quite often, the fermentation vats are kept next to animal facilities or made from improvised materials (such as wood, rubber tires, etc.). Thus, the standardization of the final product is not achieved, and the sale of *cachaça* for more demanding/rigorous markets like Europe or the US is greatly impaired. Thereby, government agencies, research laboratories, and private companies together, and some individual initiatives are trying to change the *cachaça* production outlook in order to not only improve the quality of the final product but also regulate unlicensed producers.

2.5.4. Distillation

After fermentation of sugarcane juice, the medium is taken to steel distillation columns (industrial *cachaça*) or special copper-made distillers called "alembics" (majority of the *cachaça* producers). In this last process, the distillation product can be divided into three parts: (i) head or "strong water" (5%–10%), (ii) heart (approximately 80%), and (iii) tail or "weak water" (5%–20%) [19]. The head is the first distilled portion of a fermented must, and it is rich in substances such as aldehydes, methanol, and esters. The head portion has alcohol content of 65/70% v/v. The next distillate volume is called the heart and displays ethanol concentration ranging from

35/55% v/v. This fraction has lower content of chemical contaminants that affect negatively the sensory characteristics of the product. The last fraction is known as tail, where several acids and furfurals are found, presenting an alcohol content around 14% v/v. The best quality *cachaça* are produced only with the heart fraction of the distillate, and the remainder is discarded or used to produce other products such as liqueurs or bioethanol for fuel [93].

2.5.5. Aging

The *cachaça* aging process includes the storage of the distillation product in wood barrels under specific conditions (temperature, humidity, aeration, etc.) for a period not less than 1 year. Numerous biochemical reactions occur during the aging process, the main being oxidation and esterification reactions [94]:

$$R-COH(\text{alcohol}) + O_2 \rightarrow R-HCO(\text{aldehyde})(\text{oxidation reaction})$$
$$R-HCO(\text{aldehyde}) + O_2 \rightarrow R-COOH(\text{acid})(\text{oxidation reaction})$$
$$R-HCO\ (\text{aldehyde}) + R-COOH(\text{acid}) \rightarrow R-COO-R(\text{ester})\ (\text{esterification reaction})$$

Alcohols are relatively stable to oxidation but can form significant amounts of aldehydes in the presence of phenol and water. Aldehydes are highly reactive and may oxidize to form the corresponding organic acid. Through esterification reactions, acids react with alcohols to form esters, which soften the odor of aldehydes, giving a pleasant odor to the *cachaça*. Beside the aldehydes, some sulfur compounds also decrease during maturation, such as sulfide and disulfide. In addition, alcohol and water, through capillary penetration and osmosis, pass through the interstices of the timber hydrolysing the hemicellulose and lignin [95]. The hydrolysis products are extracted, enriching the distillate and enhancing the quality of the drink [93]. Sensory gains of the beverage varies according to the chemical composition of wood, the aging time, the capacity of the barrel, the porosity, and the thickness of the timber [96]. Some studies have shown that blending aged with fresh *cachaça* is feasible and reasonable since the final product has better organoleptic and commercial characteristics. Many types of woods can be used for the manufacture of high-quality aging barrels, such as amburana (*Amburana cearencis*), jequitibá (*Cariniana estrellensis* and *Cariniana legalis*), ipê (*Tabebuia* genus), balsam wood (*Myrocarpus frondosus avium* and *Prunus cerasus*)—Brazilian trees, and oak (*Quercus* genus) [97].

3. Yeast and LAB new potential applications

South America presents a wide variety of fermented and distilled beverages, which have several unique characteristics, greatly influenced by the fermentative metabolism of microorganisms. Therefore, those microorganisms present a large potential for utilization in the development of new beverages, or even in new biotechnological applications. In this context, several scientific works have focused in the isolation and characterization of such microorganisms [11, 86, 87].

3.1. Wild yeast

During fermentation, yeast and LAB cells are submitted to several stress factors, such as: high osmotic pressure and hydrostatic pressure, high concentrations of ethanol, anaerobic atmosphere, temperature, and nutrient limitation [98]. Such pressures promote the genetic adaptation of the individuals, leading to the survival of only the fittest cells. The increasing number of such alterations will lead to changes in the fermentation subproducts, some of which contribute to the organoleptic properties of the final products. Consequently, some of those subproducts may contribute to improve the beverages and, in this way, increasing the diversification of this industrial niche. Furthermore, the utilization of microorganisms isolated from traditional products, as *chicha* or *cachaça,* in the fermentation or maturation of new drinks production can lead to development of new promising products.

Recently, wild yeast isolates from *cachaça* fermentation vats in Brazil were innovatively evaluated in the production of beer [99]. First, in that study, 21 isolates belonging to the collection of the Laboratory of Cell and Molecular Biology /UFOP were surveyed for the production of aroma compounds in the beers. For that, compounds known for their influence in aroma and flavor, such as higher alcohols, esters, acetaldehyde, diacetyl, and ethanol, were analyzed by gas chromatography. After a careful analysis of each aromatic profile, two superior isolates were chosen (LBCM78 and LBCM45). In the same study, LBCM78 and LBCM45 were shown to be suitable to the production of ale and lager beers, respectively. The physicochemical composition of the produced beers were analyzed and compared to similar beers produced by commercial brewer yeast strains [99]. As a result, in the production of a wheat beer, the commercial strain WB-06 showed higher values of n-propanol than LBCM78. However, no significant differences were observed in the isobutanol and isoamyl alcohol levels. Similarly, the LBCM45 and the commercial strain W-34/70 were used to produce a lager beer, and the final products were analyzed as before. Between those two beers, the productions of isobutanol or isoamyl alcohol were similar. Moreover, no differences were observed in the ethyl acetate and diacetyl content of all the different beers produced by the four strains. As a final step of that study, beers were submitted to a sensorial evaluation by a group of trained tasters, from the Craft Brewers Association of Minas Gerais. The wild yeast showed a similar production of ethanol to the commercial strains (3.41–4.80% v/v), and the sensory analysis of the beers produced from LBCM45 and LBCM78 strains showed good acceptance in the evaluation panel [99]. These results suggest that yeast strains isolated from *cachaça* vats have a great potential for the production of new beers since a good production of volatile compounds and ethanol were observed.

3.2. Mixed fermentation yeast/LAB

Recently, our research group started a work to approach the utilization of both yeast and LAB in the fermentation of *cachaça* and beer [100]. In that study, it was verified that the presence of LAB in the fermentation of such beverages appears to contribute to the production of organic acids, which may contribute to the formation of aromatic compounds, and to the reduction of the pH, inhibiting the growth of spoiling microorganisms. In a recent collaboration between our group and a known Minas Gerais craft brewery, several LAB were isolated from beers with high acidity level. These isolates were characterized molecularly and identified as *Lactobacillus brevis*, which are mandatory hetero-fermentative microorganisms that produce lactate, ethanol, or acetic acid and CO_2 as final metabolites [40]. Moreover, in the next step of that study, those isolates were tested in fermentations at different temperatures in order to assess their applicability in the production of certain beer styles. The selected isolate LBCM718 showed a good growth in temperatures above 18°C, and then it was tested in beer mixed fermentations with two ale yeast strains, LBCM78 and WB-06. The viability of both yeast and LAB as well as the wort final pH were analyzed [100]. The ethanol production was not affected by the presence of *L. brevis*. Such fermentations are frequent in beers from the lambic or fruit beer style, where LAB contribute with acidity and lactate, yielding ethyl lactate—an important aromatic compound. Moreover, a test was conducted to test the LAB resistance to iso-α-acids, in the concentration range generally found in beer—17 to 55 ppm [101]. There was no formation of inhibition zones, which suggest that isolate can be used in the production of beers.

In a study from another research group, a mixed fermentation of *S. cerevisiae* and *Lactococcus lactis* led to the improvement of *cachaça* quality. When the concentrations of higher alcohols, such as propanol, isobutanol, and isoamyl alcohol from three fractions (head, tail, and heart), were compared, the researchers found higher ethanol levels in the *cachaça* from mixed fermentation than in the *cachaça* produced from a pure *S. cerevisiae* inoculum. Finally, the evaluation comparing the two products was performed by 40 trained tasters, men and women aged between 22 and 50 years old. In that evaluation, both *cachaças* were evaluated for flavor, color, and overall acceptability. The *cachaça* produced by mixed culture obtained higher scores in the categories aroma and appearance, while the *cachaça* produced from pure culture yeast showed higher global acceptance. Both beverages showed similar flavor [102].

Both studies show that the use of bacteria and yeast simultaneously in fermentation apparently does affect the growth of both cultures. Similarly, the ethanol production in these mixed fermentations was the same. Furthermore, the use of mixed fermentations appears to improve the aroma of both beer and *cachaça*, a potential alternative to the development of new products.

3.3. New spirits

Brazil is the country with the world's largest fruit production; however, there is a huge postharvest waste of raw material that generates losses to the farmer. Therefore, there is the necessity to develop new processes and products to reduce these losses. In this context, an alternative is the use of these fruits for the production of alcoholic beverages [103].

In a previous study, a research group developed a fermentation process from *cajá* (*Spondias mombin*) pulp for the production of a new beverage. In that study, *cajá* pulp was inoculated with *S. cerevisiae* and then fermented at 22°C during 10 days. Analyses of amount of alcohol and higher alcohols were carried out to determine the compounds present in the beverage. Simultaneously, the final product was subject to sensory analysis. The amount of alcohol found in the *cajá* beverage was in averaged 12°GL, comparable to those found in wines. The total amounts of higher alcohols found were about 0.7 g/L, while the values for these in wines ranging from 0.1 to 0.3 g/L [104]. Sensory analysis showed a good acceptance by the tasters [103].

In another study, it was evaluated the quality of fruit spirits produced through different treatments [105]. Mango, grape, and passion fruit were used as raw materials, and the fermentation was performed using *S. cerevisiae* cultures. Distillation was performed in copper still with controlled temperature between 85°C and 90°C, and the amount of alcohol in beverage was standardized at 40°GL. After that, oak chips and umburana chips were added to the spirits for 60, 90, and 120 days. At the end of this period, the samples were conducted to sensory analysis using 10 trained panelists, using the quantitative descriptive analysis method. All three products obtained were well accepted by the tasters, being the passion fruit distillate the best evaluated. Additionally, 90 days was the best period of aging for those particular spirits [105].

As noted in these studies, alcoholic beverages obtained from tropical fruits were well accepted in sensory tests, demonstrating the potential application of these substrates in the production of new beverages.

Another study had as objective to obtain and characterize a new spirit from the fermentation of cheese whey. The cheese whey is a by-product of the dairy industry that has a high impact in the environment. The researchers used the yeast *Kluyveromyces fragilis*, due to its ability to grow in medium containing lactose, in high yields, and without the production of toxins. In order to achieve that objective, the whey powder was acidified and deproteinized, and the resulting supernatant was used for fermentation [106]. The fermentation of the whey with a high concentration of lactose (200 g/L), after 92 hours, obtained a final product with an ethanol content of 9.6% (v/v). After distillation of the fermented beverage, the heart fraction was diluted to 40% (v/v) ethanol content. The chemical analysis revealed that the higher alcohols were the most abundant group of volatile compounds present in this fraction, containing isoamyl, isopentyl, isobutyl, and 1-propanol, all present in large quantities. Among the esters, the ethyl acetate was found the highest concentration. This compound has a significant effect on the organoleptic characteristics of wines and spirits. Furthermore, the authors concluded that it was possible to obtain a spirit with pleasant smell and taste from cheese whey, containing high concentration of lactose, this being an alternative to by-product of dairy industry [106].

From these studies, we can see distinct possibilities for the production of new beverages, by changing the yeast strain/species, or using blends of different microorganisms, such as yeast and LAB. Moreover, it is possible to use several different substrates for the production of these beverages, such as fruit and cheese whey.

4. Conclusion

Studies on South American beverages are scarce when compared to other beverages like wine, beer, or even sake. This is mainly due to years of neglect to research in these countries. Until recently, the economic difficulties of the South American countries prevented investments in scientific research. Nowadays, with the economic stability, these countries increased the scientific funding, and a new reality seems to arise. In this context, the understanding of the microorganisms present in typical South American beverages opens the door to the development of new technologies, contributing to the overall scientific and economic development of such countries. For example, the isolation of yeast in *cachaça* fermentation vats may lead to the discovery of new strains resistant to different stresses, which can be used not only to produce *cachaça,* but also to produce bioethanol. Moreover, lactic acid bacteria can promote the appearance of new products such as beers with unique or regional flavors. Thus, studies of these microorganisms diversity, present in such unique environments as the traditional beverages, help uncovering new potential applications. Furthermore, the knowledge of those microorganisms can promote the revival of traditional beverages, as *cauim* that was the most consumed beverage in South America in the centuries XVI–XVII. Nevertheless, despite the increased investment in research, few laboratories have the know-how necessary or the availability of resources to invest in the "screening" of yeast or lactic acid bacteria in traditional beverages. Research on South American traditional beverages is important to improve the quality of those beverages, but also to develop new products from these microorganisms.

Acknowledgements

The authors were supported by grants from the following Brazilian agencies: Conselho Nacional de Desenvolvimento Científico e Tecnológico (CNPq), Coordenação de Aperfeiçoamento de Pessoal de Nível Superior (CAPES), and Fundação de Amparo à Pesquisa do Estado de Minas Gerais.

Author details

Fábio Faria-Oliveira, Raphael H.S. Diniz, Fernanda Godoy-Santos, Fernanda B. Piló, Hygor Mezadri, Ieso M. Castro and Rogelio L. Brandão*

*Address all correspondence to: rlbrand@ufop.br

Laboratório de Biologia Celular e Molecular, Núcleo de Pesquisas em Ciências Biológicas, Universidade Federal de Ouro Preto, Brazil

References

[1] McGovern PE, Zhang J, Tang J, Zhang Z, Hall GR, Moreau RA, Nuñez A, Butrym ED, Richards MP, Wang C. Fermented beverages of pre- and proto-historic China. Proc Natl Acad Sci USA. 2004;101:17593.

[2] Faria-Oliveira F, Puga S, Ferreira C. Yeast: World's finest *Chef.* In: Muzzalupo I, editor. Food Industry. Rijeka: Intech; 2013. p. 519–547. DOI: 10.5772/53156

[3] Erten H, Ağirman B, Gündüz C, Çarşanba E, Sert S, Bircan S, Tangüler H. Importance of yeast and lactic acid bacteria in food processing. In: Malik A, Erginkaya Z, Ahmad S, Erten H, editors. Food Processing: Strategies for Quality Assessment. New York: Springer; 2014. p. 351–378. DOI: 10.1007/978-1-4939-1378-7_14

[4] Katina K, Poutanen K. Nutritional aspects of cereal fermentation with lactic acid bacteria and yeast. In: Gobbetti M, Gänzle M, editors. Handbook on Sourdough Biotechnology. Springer US; 2013. p. 229–244. DOI: 10.1007/978-1-4614-5425-0_9

[5] Larsson K, Ansell R, Eriksson P, Adler L. A gene encoding a glycerol 3-phosphate dehydrogenase (NAD+) complements an osmosensitive mutant of *Saccharomyces cerevisiae*. Mol Microbiol. 1993;10:1101–1111.

[6] Kurtzman CP, Fell JW, Boekhout T. The Yeast: A Taxonomic Study. Amsterdam: Elsevier; 2011.

[7] Kejžar A, Gobec S, Plemenitaš A, Lenassi M. Melanin is crucial for growth of the black yeast *Hortaea werneckii* in its natural hypersaline environment. Fungal Biol. 2013;117:368–379. DOI: 10.1016/j.funbio.2013.03.006

[8] Gadanho M, Libkind D, Sampaio JP. Yeast diversity in the extreme acidic environments of the Iberian pyrite belt. Microbial Ecol. 2006;52:552–563. DOI: 10.1007/s00248-006-9027-y

[9] Hashim N, Bharudin I, Nguong D, Higa S, Bakar F, Nathan S, Rabu A, Kawahara H, Illias R, Najimudin N, Mahadi N, Murad A. Characterization of Afp1, an antifreeze protein from the psychrophilic yeast *Glaciozyma antarctica PI12*. Extremophiles. 2013;17:63–73. DOI: 10.1007/s00792-012-0494-4

[10] Tsuji M, Yokota Y, Shimohara K, Kudoh S, Hoshino T. An application of wastewater treatment in a cold environment and stable lipase production of antarctic basidiomycetous yeast *Mrakia blollopis*. PLoS One. 2013;8:e59376. DOI: 10.1371/journal.pone.0059376

[11] Conceição LEFR, Saraiva MAF, Diniz RHS, Oliveira J, Barbosa GD, Alvarez F, da Mata Correa LF, Mezadri H, Coutrim MX, Afonso RJdCF, Lucas C, Castro IM, Brandão RL. Biotechnological potential of yeast isolates from cachaça: the Brazilian spirit. ""Journal of Industrial Microbiology and Biotechnology. 2015;42:237–246. DOI: 10.1007/s10295-014-1528-y

[12] Steensels J, Verstrepen KJ. Taming wild yeast: potential of conventional and nonconventional yeast in industrial fermentations. Ann Rev Microbiol. 2014;68:61–80. DOI: 10.1146/annurev-micro-091213-113025

[13] Kongkiattikajorn J. Production of amylase from *Saccharomyces diastaticus* sp. and hydrolysis of cassava pulps for alcohol production. J Agric Sci Technol B. 2012;2:909–918.

[14] Laluce C, Mattoon JR. Development of rapidly fermenting strains of *Saccharomyces diastaticus* for direct conversion of starch and dextrins to ethanol. App Environ Microbiol. 1984;48:17–25.

[15] Spencer-Martins I. Transport of sugars in yeast: implications in the fermentation of lignocellulosic materials. Biores Technol. 1994;50:51–57.

[16] Kruckeberg AL, Dickinson JR. Carbon metabolism. In: Dickinson JR, Schweizer M, editors. The metabolism and molecular physiology of *Saccharomyces cerevisiae*. London: CRC; 2004. p. 42–103.

[17] Gancedo JM. The early steps of glucose signalling in yeast. FEMS Microbiol Rev. 2008;32:673–704.

[18] Galdieri L, Mehrotra S, Yu S, Vancura A. Transcriptional regulation in yeast during diauxic shift and stationary phase. Omics. 2010;14:629–638.

[19] Meijer MM, Boonstra J, Verkleij AJ, Verrips CT. Glucose repression in *Saccharomyces cerevisiae* is related to the glucose concentration rather than the glucose flux. J Biol Chem. 1998;273:24102–24107.

[20] Rintala E, Wiebe M, Tamminen A, Ruohonen L, Penttilä M. Transcription of hexose transporters of *Saccharomyces cerevisiae* is affected by change in oxygen provision. BMC Microbiol. 2008;8:53.

[21] Nelson DL, Cox MM. Lehninger Principles of Biochemistry. New York: W. H. Freeman; 2008.

[22] Rodrigues F, Ludovico P, Leão C. Sugar metabolism in yeast: an overview of aerobic and anaerobic glucose catabolism. In: Rosa CA, Peter G, editors. Biodiversity and Ecophysiology of Yeast. 2006. p. 101–121.

[23] Pretorius IS. Tailoring wine yeast for the new millennium: novel approaches to the ancient art of winemaking. Yeast. 2000;16:675–729.

[24] Pronk JT, Steensma HY, Van Dijken JP. Pyruvate metabolism in *Saccharomyces cerevisiae*. Yeast. 1996;12:1607–1633.

[25] van Dijken JP, Scheffers WA. Redox balances in the metabolism of sugars by yeast. FEMS Microbiol Lett. 1986;32:199–224.

[26] Eriksson P, André L, Ansell R, Blomberg A, Adler L. Cloning and characterization of *GPD2*, a second gene encoding a *DL*-glycerol 3-phosphate dehydrogenase (NAD+) in *Saccharomyces cerevisiae*, and its comparison with *GPD1*. Mol Microbiol. 1995;17:95–107.

[27] Norbeck J, Påhlman AK, Akhtar N, Blomberg A, Adler L. Purification and characterization of two isoenzymes of *DL*-glycerol-3-phosphatase from *Saccharomyces cerevisiae*. J Biol Chem. 1996;271:13875–13881.

[28] Crabtree HG. Observations on the carbohydrate metabolism of tumours. Biochem J. 1929;23:536.

[29] De Deken R. The Crabtree effect: a regulatory system in yeast. J Gen Microbiol. 1966;44:149.

[30] Golshani-Hebroni SG, Bessman SP. Hexokinase binding to mitochondria: a basis for proliferative energy metabolism. J Bioenerg Biomemb. 1997;29:331–338.

[31] Skinner C, Lin SJ. Effects of calorie restriction on life span of microorganisms. Appl Microbiol Biotechnol. 2010;88:817–828.

[32] Murray DB, Haynes K, Tomita M. Redox regulation in respiring *Saccharomyces cerevisiae*. Biochim Biophys Acta. 2011;1810:945–958.

[33] Feldmann H. Yeast metabolism. In: Feldmann H, editor. Yeast Molecular Biology—A Short Compendium on Basic Features and Novel Aspects. Munich: Adolf-Butenandt-Institute, University of Munich; 2005.

[34] Boone DR, Castenholz RW, Garrity GM, Brenner DJ, Krieg NR, Staley JT. Bergey's Manual® of Systematic Bacteriology. Springer Science & Business Media; 2005.

[35] Liu W, Pang H, Zhang H, Cai Y. Biodiversity of Lactic Acid Bacteria. In: Zhang H, Cai Y, editors. Lactic Acid Bacteria. Springer Netherlands; 2014. p. 103–203. 10.1007/978-94-017-8841-0_2

[36] Hoover DG, Steenson LR. Bacteriocins of lactic acid bacteria. Academic Press; 2014.

[37] Horvath P, Coute-Monvoisin AC, Romero DA, Boyaval P, Fremaux C, Barrangou R. Comparative analysis of *CRISPR* loci in lactic acid bacteria genomes. Int J Food Microbiol. 2009;131:62–70. DOI: 10.1016/j.ijfoodmicro.2008.05.030

[38] Stiles ME, Holzapfel WH. Lactic acid bacteria of foods and their current taxonomy. Int J Food Microbiol. 1997;36:1–29.

[39] Klein G, Pack A, Bonaparte C, Reuter G. Taxonomy and physiology of probiotic lactic acid bacteria. Int J Food Microbiol. 1998;41:103–125.

[40] Felis GE, Dellaglio F. Taxonomy of lactobacilli and bifidobacteria. Curr Issues Intest Microbiol. 2007;8:44.

[41] Giraffa G, Chanishvili N, Widyastuti Y. Importance of lactobacilli in food and feed biotechnology. Res Microbiol. 2010;161:480–487. DOI: 10.1016/j.resmic.2010.03.001

[42] König H, Fröhlich J. Lactic acid bacteria. In: Biology of Microorganisms on Grapes, in Must and in Wine. Springer; 2009. p. 3–29.

[43] Ananou S, Maqueda M, Martínez-Bueno M, Valdivia E. Biopreservation, an ecological approach to improve the safety and shelf-life of foods. In: Communicating Current Research and Educational Topics and Trends in Applied Microbiology. Formatex; 2007. p. 475–486.

[44] Howarth GS, Wang H. Role of endogenous microbiota, probiotics and their biological products in human health. Nutrients. 2013;5:58–81.

[45] White D, Drummond JT, Fuqua C. The physiology and biochemistry of prokaryotes. New York: Oxford University Press; 2007.

[46] Romano AH, Eberhard SJ, Dingle SL, McDowell TD. Distribution of the phosphoenolpyruvate: glucose phosphotransferase system in bacteria. J Bacteriol. 1970;104:808–813.

[47] Kandler O. Carbohydrate metabolism in lactic acid bacteria. Antonie van Leeuwenhoek. 1983;49:209–224.

[48] Leroy F, De Vuyst L. Lactic acid bacteria as functional starter cultures for the food fermentation industry. Trends Food Sci Tech. 2004;15:67–78.

[49] Justé A, Malfliet S, Waud M, Crauwels S, De Cooman L, Aerts G, Marsh TL, Ruyters S, Willems K, Busschaert P. Bacterial community dynamics during industrial malting, with an emphasis on lactic acid bacteria. Food Microbiol. 2014;39:39–46.

[50] Bokulich NA, Bamforth CW. The microbiology of malting and brewing. Microbiol Mol Biol Rev. 2013;77:157–172. DOI: 10.1128/MMBR.00060-12

[51] Smid EJ, Kleerebezem M. Production of aroma compounds in lactic fermentations. Annu Rev Food Sci Technol. 2014;5:313–326. DOI: 10.1146/annurev-food-030713-092339

[52] Soomro AH, Masud T, Kiran A. Role of lactic acid bacteria (LAB) in food preservation and human health–a review. Pak J Nutr. 2002;

[53] Liu SQ. A review: malolactic fermentation in wine—beyond deacidification. J Appl Microbiol. 2002;92:589–601.

[54] Swiegers JH, Bartowsky EJ, Henschke PA, Pretorius IS. Yeast and bacterial modulation of wine aroma and flavour. Aust J Grape Wine Res. 2005;11:139–173.

[55] Nedovic VA, Durieuxb A, Van Nedervelde L, Rosseels P, Vandegans J, Plaisant AM, Simon JP. Continuous cider fermentation with co-immobilized yeast and *Leuconostoc oenos* cells. Enzyme Microb Tech. 2000;26:834–839.

[56] Rouse S, Sun F, Vaughan A, Sinderen D. High-throughput isolation of bacteriocin-producing lactic acid bacteria, with potential application in the brewing industry. J Inst Brew. 2007;113:256–262.

[57] Vaughan A, Eijsink VGH, O'Sullivan TF, O'Hanlon K, Van Sinderen D. An analysis of bacteriocins produced by lactic acid bacteria isolated from malted barley. J Appl Microbiol. 2001;91:131–138.

[58] Lowe DP, Arendt EK. The use and effects of lactic acid bacteria in malting and brewing with their relationships to antifungal activity, mycotoxins and gushing: a review. J Inst Brew. 2004;110:163–180.

[59] Almeida EG, Rachid CC, Schwan RF. Microbial population present in fermented beverage 'cauim' produced by Brazilian Amerindians. Int J Food Microbiol. 2007;120:146–151. DOI: 10.1016/j.ijfoodmicro.2007.06.020

[60] Aidoo KE, Nout MJ, Sarkar PK. Occurrence and function of yeast in Asian indigenous fermented foods. FEMS Yeast Research. 2006;6:30–39. DOI: 10.1111/j.1567-1364.2005.00015.x

[61] Blandino A, Al-Aseeri M, Pandiella S, Cantero D, Webb C. Cereal-based fermented foods and beverages. Food Res Int. 2003;36:527–543.

[62] Osorio-Cadavid E, Chaves-Lopez C, Tofalo R, Paparella A, Suzzi G. Detection and identification of wild yeast in Champús, a fermented Colombian maize beverage. Food Microbiol. 2008;25:771–777. DOI: 10.1016/j.fm.2008.04.014

[63] Ramos CL, de Almeida EG, Pereira GVdM, Cardoso PG, Dias ES, Schwan RF. Determination of dynamic characteristics of microbiota in a fermented beverage produced by Brazilian Amerindians using culture-dependent and culture-independent methods. Int J Food Microbiol. 2010;140:225–231. DOI: 10.1016/j.ijfoodmicro.2010.03.029

[64] Santos CC, Almeida EG, Melo GV, Schwan RF. Microbiological and physicochemical characterisation of *caxiri*, an alcoholic beverage produced by the indigenous *Juruna* people of Brazil. Int J Food Microbiol. 2012;156:112–121. DOI: 10.1016/j.ijfoodmicro.2012.03.010

[65] Steinkraus K. Handbook of Indigenous Fermented Foods, revised and expanded. CRC Press; 1995.

[66] Vallejo JA, Miranda P, Flores-Félix JD, Sánchez-Juanes F, Ageitos JM, González-Buitrago JM, Velázquez E, Villa TG. Atypical yeast identified as *Saccharomyces cerevisiae* by MALDI-TOF MS and gene sequencing are the main responsible of fermentation of chicha, a traditional beverage from Peru. Syst Appl Microbiol. 2013;36:560–564. DOI: 10.1016/j.syapm.2013.09.002

[67] Gomes F, Lacerda I, Libkind D, Lopes C, Carvajal E, Rosa C. Traditional foods and beverages from South America: microbial communities and production strategies. In: Krause J, Fleischer O, editors. Industrial Fermentation: Food Processes, Nutrient Sources and Production Strategies. 2009. p. 79–114.

[68] Piló FB. Leveduras e bactérias lácticas associadas à *chicha*, uma bebida tradicional produzida no Equador [thesis]. Belo Horizonte, MG: Universidade Federal de Minas Gerais; 2014.

[69] Elizaquível P, Pérez-Cataluña A, Yépez A, Aristimuño C, Jiménez E, Cocconcelli PS, Vignolo G, Aznar R. Pyrosequencing vs. culture-dependent approaches to analyze

lactic acid bacteria associated to *chicha*, a traditional maize-based fermented beverage from Northwestern Argentina. Int J Food Microbiol. 2015;198:9–18.

[70] Lago C, Landoni M, Cassani E, Doria E, Nielsen E, Pilu R. Study and characterization of a novel functional food: purple popcorn. Mol Breed. 2013;31:575–585.

[71] Schwarz M, Hillebrand S, Habben S, Degenhardt A, Winterhalter P. Application of high-speed countercurrent chromatography to the large-scale isolation of anthocyanins. Biochem Eng J. 2003;14:179–189. DOI: 10.1016/S1369-703X(02)00219-X

[72] Miller C. Fruit production of the ungurahua palm (*Oenocarpus bataua* subsp. *bataua*, Arecaceae) in an indigenous managed reserve. Econ Bot. 2002;56:165–176.

[73] Cox L, Caicedo B, Vanos V, Heck E, Hofstaetter S, Cordier J. A catalogue of some ecuadorean fermented beverages, with notes on their microflora. World J Microbiol Biotechnol. 1987;3:143–153.

[74] Puerari C, Magalhães-Guedes KT, Schwan RF. Physicochemical and microbiological characterization of chicha, a rice-based fermented beverage produced by Umutina Brazilian Amerindians. Food Microbiol. 2015;46:210–217.

[75] Carvajal Barriga EJ. Arqueología Microbiana. Nuestra Ciencia. 2012;14:3–7.

[76] Chang CF, Lin YC, Chen SF, Carvajal Barriga EJ, Barahona PP, James SA, Bond CJ, Roberts IN, Lee CF. *Candida theae* sp. nov., a new anamorphic beverage-associated member of the *Lodderomyces* clade. Int J Food Microbiol. 2012;153:10–14. DOI: 10.1016/j.ijfoodmicro.2011.09.012

[77] Cavalcante MS. A verdadeira história da cachaça. São Paulo: Sá editora; 2011. 608 p.

[78] Lei 10958/04. Presidency of the Republic. Brasilia. Brazil, 2004.

[79] IBRAC. Instituto Brasileiro da Cachaça. IBRAC. 03/08/2015. http://www.ibrac.net/

[80] Lobo P, Jaguaribe E, Rodrigues J, Da Rocha F. Economics of alternative sugar cane milling options. Appl Therm Eng. 2007;27:1405-1413.

[81] Meade GP, Chen JC. Cane Sugar Handbook. New York: John Wiley & Sons; 1977.

[82] Duarte WF, Amorim JC, Schwan RF. The effects of co-culturing non-*Saccharomyces* yeast with *S. cerevisiae* on the sugar cane spirit (cachaça) fermentation process. Antonie van Leeuwenhoek. 2013;103:175–194.

[83] Aquarone E, Borzani W, Schmidell W, Lima UdA. Biotecnologia Industrial. São Paulo: Edgard Blücher Ltda; 2001. 523 p.

[84] Passoth V, Olstorpe M, Schnurer J. Past, present and future research directions with *Pichia anomala*. Antonie van Leeuwenhoek. 2011;99:121–125. DOI: 10.1007/s10482-010-9508-3

[85] Gomes F, Silva C, Marini M, Oliveira E, Rosa C. Use of selected indigenous *Saccharomyces cerevisiae* strains for the production of the traditional cachaça in Brazil. J Appl Microbiol. 2007;103:2438-2447.

[86] Vicente MA, Fietto LG, Castro IM, dos Santos AN, Coutrim MX, Brandao RL. Isolation of *Saccharomyces cerevisiae* strains producing higher levels of flavoring compounds for production of "*cachaça*" the Brazilian sugarcane spirit. Int J Food Microbiol. 2006;108:51–59. DOI: 10.1016/j.ijfoodmicro.2005.10.018

[87] Oliveira VA, Vicente MA, Fietto LG, Castro IM, Coutrim MX, Schuller D, Alves H, Casal M, Santos JO, Araujo LD, da Silva PH, Brandao RL. Biochemical and molecular characterization of *Saccharomyces cerevisiae* strains obtained from sugar-cane juice fermentations and their impact in *cachaça* production. Appl Environ Microbiology. 2008;74:693–701. DOI: 10.1128/AEM.01729-07

[88] Campos C, Silva C, Dias D, Basso L, Amorim H, Schwan R. Features of *Saccharomyces cerevisiae* as a culture starter for the production of the distilled sugar cane beverage, cachaça in Brazil. J Appl Microbiol. 2010;108:1871–1879.

[89] de Souza APG, Vicente MdA, Klein RC, Fietto LG, Coutrim MX, de Cássia Franco Afonso RJ, Araújo LD, da Silva PHA, Bouillet LÉM, Castro IM, Brandão RL. Strategies to select yeast starters cultures for production of flavor compounds in *cachaça* fermentations. Antonie van Leeuwenhoek. 2012;101:379–392. DOI: 10.1007/s10482-011-9643-5

[90] Schwan RF, Mendonca AT, da Silva Jr JJ, Rodrigues V, Wheals AE. Microbiology and physiology of *Cachaça* (*Aguardente*) fermentations. Antonie van Leeuwenhoek. 2001;79:89–96.

[91] Oliveira Cd, Garíglio H, Ribeiro M, Alvarenga M, Maia F. Cachaça de Alambique–Manual de Boas Práticas Ambientais e de Produção. Belo Horizonte. SEAPA/SEMAD/AMPAQ/FEAM/IMA, 2005.

[92] Martino DBd. Aguardente: O destilado do século 21. Revista Engarrafador Moderno. 1998;84–88.

[93] Cardeal ZL, de Souza PP, da Silva MD, Marriott PJ. Comprehensive two-dimensional gas chromatography for fingerprint pattern recognition in *cachaça* production. Talanta. 2008;74:793–799. DOI: 10.1016/j.talanta.2007.07.021

[94] Bortoletto AM, Alcarde AR. Assessment of chemical quality of Brazilian sugar cane spirits and cachaças. Food Control. 2015;54:1–6.

[95] da Silva AA, do Nascimento ES, Cardoso DR, Franco DW. Coumarins and phenolic fingerprints of oak and Brazilian woods extracted by sugarcane spirit. J Sep Sci. 2009;32:3681–3691. DOI: 10.1002/jssc.200900306

[96] Bortoletto AM, Alcarde AR. Congeners in sugar cane spirits aged in casks of different woods. Food Chem. 2013;139:695–701. DOI: 10.1016/j.foodchem.2012.12.053

[97] de Souza PP, Siebald HG, Augusti DV, Neto WB, Amorim VM, Catharino RR, Eberlin MN, Augusti R. Electrospray ionization mass spectrometry fingerprinting of Brazilian artisan *cachaça* aged in different wood casks. J Agric Food Chem. 2007;55:2094–2102. DOI: 10.1021/jf062920s

[98] Lei H, Zhao H, Yu Z, Zhao M. Effects of wort gravity and nitrogen level on fermentation performance of brewer's yeast and the formation of flavor volatiles. Appl Biochem Biotechnol. 2012;166:1562–1574.

[99] Araújo TM. Caracterização bioquímico-molecular de cepas de *Saccharomyces cerevisiae* isoladas de dornas de fermentação de cachaça para produção de cervejas [thesis]. Universidade Federal de Ouro Preto; 2013.

[100] Campos ACS. Caracterização de bactérias lácticas para serem utilizadas em processos fermentativos consorciados entre leveduras e bactérias na produção de cerveja. [thesis]. Universidade Federal de Ouro Preto; 2014.

[101] Sakamoto K, Konings WN. Beer spoilage bacteria and hop resistance. Int J Food Microbiol. 2003;89:105–124.

[102] Carvalho FP, Duarte WF, Dias DR, Piccoli RH, Schwan RF. Interaction of *Saccharomyces cerevisiae* and *Lactococcus lactis* in the fermentation and quality of artisanal cachaça. Acta Sci Agron. 2014;37:51–60.

[103] Dias DR, Schwan RF, Lima LCO. Metodologia para elaboração de fermentado de cajá (*Spondias mombin L.*). Ciênc Tecnol Aliment. 2003;23:342–350.

[104] Vogt E, Jakob L, Lemperle E, Weiss E. El vino: obtención, elaboración y análisis. Zaragoza: Acribia; 1986.

[105] da Silva MC, de Azevedo LC, de Carvalho MM, de Sá AGB, dos Santos Lima M. Elaboração e avaliação da qualidade de aguardentes de frutas submetidas a diferentes tratamentos. Revista Semiárido De Visu. 2011;1:92–106.

[106] Dragone G, Mussatto SI, Vilanova M, Oliveira J, Teixeira J, Silva JBA. Obtenção e caracterização de bebida destilada a partir da fermentação do soro de queijo. Braz J Food Tech. 2009;120–124.

Chapter 5

Safety and Quality in the Agricultural Product Chain in Brazil

Elisa Y. Hirooka, Elisabete Y.S. Ono, Cássia R. Takabayashi-Yamashita, Angélica T. Ishikawa, Douglas F. Barbin, Jaqueline G. Bordini, Melissa T. Hirozawa, Wagner E. Risso, Ana L.S.M. Felício, Artur K. Bagatin, Tatiana Á. Miguel, Thiago M. Souza, Yumi Tabuchi, Josemarque L. Rosa, Leonardo F. Maciel, Dani L. D. Silva, Renata P. Sobottka, André M. Prando, Sandra Garcia, Claudemir Zucareli, Emilia K. Kuroda, Elisabete H. Hashimoto, Diva de Souza Andrade, Martha Z. Miranda, Maria S. Miranda, Eliete S. Bispo, Adriana L. Soares, Massami Shimokomaki, Myrna Sabino, Eiko N. Itano, Tomoaki Tsutsumi, Satoshi Nagata, Yoshitsugu Sugiura, Ken-Ichi Harada and Osamu Kawamura

Additional information is available at the end of the chapter

http://dx.doi.org/10.5772/60593

Abstract

An agriculture-intensive country should be aware of natural toxins, including both mycotoxins and cyanotoxins, which are closely associated with the quality of raw materials, for food safety and industry. The major production chains – corn, wheat, beef, and broiler chicken – are the top components of agribusiness, and they should be tracked by reliable and practical tools. The corn chain is of particular concern in food production; intensive controls, multi-year mycotoxin monitoring, and improved harmless/sustainable management methods for uninterrupted farming in the tropic-subtropics are needed to achieve a long-lasting trend. The rapid control of natural toxins (mycotoxin and cyanotoxin) has focused on immunochemical methods developed with highly specific monoclonal antibodies (mAb) matched with chromatographic methods. In parallel, the promising widespread application of non-destructive analytical methods based on NIR (Near Infrared Reflectance) spectroscopy, computer vision and hyperspectral imaging coupled with multivariate analyses have been introduced as an alternative for the prediction of quality and compositional parameters. Rapid quality control and product traceability are discussed, as well as accurate monitoring, which is essential for potentially launching an innovative system for food production in Brazil.

Keywords: Food quality and safety, rapid methods, immunoassay, natural toxins, sustainability

1. Introduction

Food matrices are organic materials with varied compositions, which also provide the nutritional components and perfect growing conditions for microorganisms, coupled with the simultaneous occurrence of several metabolic activities. Agricultural products, such as vegetables and fruits, are the basis of the food chain. Additional mechanical injuries during postharvest processing, storage and transportation may cause further points of contamination, leading to reduced quality, and compromising safety.

Preserving food products involves controlling external and internal conditions to avoid undesirable microbial growth and/or degradation processes, as well as the biosynthesis of unavoidable secondary metabolites, namely mycotoxins and phycotoxins.

Globalization demands high quality and competitiveness throughout the food chain. Quality and safety are typically achieved through a Hazard Analysis and Critical Control Points (HACCP) Risk Assessment. Providing raw materials of high quality and safe ingredients also includes the quality of water employed in food processes, which should be free of contaminants. Such strategies involve the detection of toxic secondary metabolites through continuous monitoring with reliable analytical methods, which should not only be restricted at the qualitative occurrence level but also the exact quantitative level compared with the maximal contamination limit proposed by guidelines.

The recommended techniques to detect ng and μg levels of toxic metabolites, and waste contaminants are based on High Performance Liquid Chromatography (HPLC) coupled with high sensitivity mass spectrometry (MS), which analyses residual contamination in a wide variety of products and materials.

These improvements are in contrast with the reality in raw material producing countries, highlighting the need for the innovative implementation of rapid methods combining simplicity, sensitivity and accuracy. Additionally, sequential processing and material resources in the food industry should be continuously monitored for safety and quality, which requires rapid monitoring *in loco*.

The rapid detection of natural toxins, such as mycotoxins and cyanotoxins, has been focused on immunochemical methods developed with highly specific monoclonal antibodies (mAb)

matched with chromatographic methods. Such techniques arose based on antibodies, highlighting the immunoaffinity column (IAC) for the clean-up step, and enzyme linked immunosorbent assay (ELISA) with the advantage of eliminating toxic solvents (using buffer). The current commercial kits have been the practical tool of choice and have an important role in avoiding hazards for animals and humans. Immunoassays are advancing with developments in nano-engineering, resulting in compact, miniaturized electronic devices, such as biosensors, which combine high specificity and biological diversity with automation of diagnostics. The advantages of these developments are their specificity, speed and simplicity for the detection of dangerous levels of natural toxins.

Nevertheless, both chemical and biological analytical methods are destructive, i.e., the decision concerning the total batch is extrapolated based on data obtained with samples that were already destroyed for analysis. Non-invasive and non-destructive chemical-free techniques came as a welcome option in the industrial process, including optical methods (Fourier Transform coupled to Infrared Spectroscopy, FT-IR spectroscopy and transmittance in the near infrared, near-IR), as well as an "electronic nose" for volatile compounds. These technologies are able to integrate with online quality control monitoring systems in the food chain in real time and are able to detect imbalances caused by deteriorated quality, which can also indicate undesirable toxic metabolites.

Continuous tracking in the food chain should focus on safety and quality using practical and reliable analytical techniques. The combination of rapid biological assays, non-destructive physical technologies and primary chemical analysis are desirable procedures for extending the shelf life of a product.

We begin by presenting data on corn, a topic of concern in the food chain, as it is a universal ingredient with unavoidable mycotoxin hazards – even with extensive monitoring in the agro-industrial region of Southern Brazil. An ic-ELISA-based immunoassay was developed, established and optimized to analyse different food groups, using specific MAb produced by hybridomas (especially against non-immunogenic low molecular mass ochratoxin (OTA), aflatoxin (AF), deoxynivalenol (DON), zearalenone (ZEA), fumonisin B_1(FB_1), OTA and microcystin-LR (MCLR)). This has become important for rapid tracking, monitoring safety and quality, and providing guidance for the best conduct to establish a long-lasting trend focusing on harmless/sustainable management in uninterrupted tropical-subtropical farming, and in replacing chemical agrotoxicants. The control of natural toxins should begin at the field level through sustainable management, adequate water quality, predictive modelling, as well as in the food processing systems in agroindustry. Such an overall approach, could result in the production of healthy foods in potential food producing regions in Brazil.

2. Corn chain – Relevant aspects in a producing country

Brazil has a vast cultivated area of 53.20 million ha and continues to expand production, with 10.9 % in grain volume corresponding to 184.30 million from the 2012/2013 crop harvest, compared with 166.17 million ton in 2011/2012 [1]. Estimates indicate that grain production will be approximately 200.08 million ton in 2014/2015, 3.4 % higher than the 2013/2014 production [2]. The total cultivated area of grain also showed growth (57.03 million to 57.39 million hectares), with the promising increase of second crops allowed by tropical climates.

Corn (*Zea mays* L.) is one of the major cereal crops in Brazil, with an annual production of 78.689 million (metric) ton, ranking the country as the third largest corn producer in the world. Paraná State was the second largest producer with 14.504 million ton in 2014 [3]. An abrupt increase in trading has occurred due to the large territorial extension associated with climatic diversity; i.e., the tropical climate in midwestern and northern regions, and subtropical-temperate climates in southern and southeastern Brazil, which enables production throughout the year. Therefore, the exportation of 3.37 million ton in February 2013 was 297 % higher than in January 2012. Corn is the second most produced crop (39.29 %), preceded only by soybean (47.19 %). From February 2014 to January 2015 – with 20.9 million ton, Brazilian exports were 400,000 ton higher than the projected capacity. The main corn importers were Iran, Vietnam, South Korea, Taiwan, Egypt, Indonesia, Malaysia, Japan, Saudi Arabia and Morocco [2].

Figure 1 shows the production of corn immediately following the soybean harvest, which allows up to three crop cycles per year in producing regions.

Figure 1. Harvesting of precocious soybean followed by simultaneous planting of corn as the second crop.

Approximately 70 % of Brazilian corn is intended for swine and broiler feeds, whereas processing for human consumption corresponds to 15 % [4]. Nevertheless, its high nutritional quality introduces risk for the growth of toxigenic fungi favoured by tropical and subtropical climates. Mycotoxins are natural thermostable metabolites responsible for substantial economic losses and their persistent residual levels can be detected even in post-processed meat, eggs, milk and dairy products. The Food and Agricultural Organization (FAO) estimated the worldwide mycotoxin contamination in crops at 25 %.

The most important mycotoxins in tropical developing countries, such as Brazil, have been fumonisins produced mainly by *Fusarium verticillioides* and *F. proliferatum* [5], and aflatoxins produced by *Aspergillus flavus* and *A. parasiticus* [6].

Corn quality associated with fungal and mycotoxin contamination in Paraná State, southern Brazil, has been studied since the 1980s, when fumonisin caused animal poisonings. The first report involving an animal outbreak detected fumonisin B_1 (FB_1) and B_2 (FB_2) in feed samples and determined that *F. verticillioides* isolates were acutely toxic to ducklings [7]. A survey of fumonisin in corn kernels also reported its occurrence in the states of Mato Grosso do Sul and Goiás [8], the most progressive region of grain production, where FB_1 and FB_2 were detected in 97.4 % and 94.8 % samples, respectively [2]. All corn from Northern Paraná (n = 39) was fumonisin positive (mean level of FB_1 = 4.79 μg g^{-1} and FB_2 = 3.95 μg g^{-1}). The samples from Mato Grosso and Goiás States (n = 9) showed mean FB_1 levels of 10.59 and 5.83 μg g^{-1} and FB_2 levels of 10.31 and 3.62 μg g^{-1}, respectively.

The co-occurrence of fumonisins and aflatoxins was investigated in 150 freshly harvested corn samples (1994/1995 crop) from the central-southern (n = 27 samples), central-western (n = 86) and northern (n = 37) regions of Parana State. Fumonisins and aflatoxins were detected in 98 % and 11.3 % samples, respectively. All the aflatoxin positive samples (mean, 191 ng g^{-1}) were from the central-western region and were co-contaminated with fumonisins. Higher fumonisin levels were detected in corn from the northern (9.85 μg g^{-1}) and central-western regions (5.08 μg g^{-1}) relative to the central-southern region (1.14 μg g^{-1}), suggesting an effect of climatic conditions in addition to the local predominance of toxigenic *Fusarium* biotypes [9].

Fumonisin monitoring in real time was established (2003-2004 crop) on critical steps (field, reception and pre-drying) of the corn chain [10]. Fumonisins were analysed in 490 samples of freshly harvested corn (2003-2004 crop) collected at three points of the production chain in Northern Paraná State, and correlated with the time interval between harvesting and the pre-drying step. The mean fumonisin level increased gradually from ≤ 5.0 μg g^{-1} to 19.0 μg g^{-1} when the time interval between harvesting and the pre-drying step increased from 3.22 to 8.89 hours. Fumonisin levels were correlated positively ($p \leq 0.05$) with time interval (ϱ=0.96), indicating that a delay in the drying process could increase the levels of contamination.

A study [11] evaluated fumonisin in 870 freshly harvested corn samples (2003 and 2004 crops) used by processing industries in Northern Paraná State. Sampling was performed at two points of the corn chain, i.e., at reception and the pre-drying step in the processing industry. Fumonisins (FB_1 + FB_2) were detected in all samples from the two points in both crops. Fumonisin levels in reception (2.24 μg g^{-1}) and pre-drying samples (2.87 μg g^{-1}) of the 2003 and 2004 crops (1.46 and 1.52 μg g^{-1}, respectively) showed similar profiles, indicating that corn used by processing industries in this region showed lower fumonisin levels than in previous studies [8, 9, 12]. Years of monitoring have shown a decreasing trend of fumonisin contamination, which may be due to changing procedures at food and feed processing facilities.

Because determination of the degree of exposure is one of the most important parameters concerning the risk assessment of chemical compounds, a study [13] estimated the maximum probable daily intake (PDI_M) of fumonisins in a local population. This study was based on fumonisin monitoring in 300 freshly harvested corn (2003 and 2004 crops) samples collected at two points of the production chain (reception and pre-drying) in Northern Paraná State. Based on the highest mean fumonisin levels being detected in the pre-drying samples (3.12 μg g^{-1}) and the average consumption of corn-based products, the maximum probable daily intake (PDI_M) of FB_1 estimated in the Brazilian population (0.95 μg kg^{-1} body weight day^{-1}) was below the tolerable daily intake (2.0 μg kg^{-1} body weight day^{-1}).

Such monitoring allowed the identification of fumonisin levels in different regions of the state, enabling it to gain a prominent position in corn exportation. Currently, the State of Paraná is responsible for 14.3 million tons/year, corresponding to 17.9 % of the national corn production [3].

3. Rapid immunoreagent monitoring in food safety

Advanced techniques in liquid chromatography using different detectors (UV-Vis-PDA, FLD, MS and LC-MS/MS) have been introduced for the analysis of chemicals in different matrices (food, microbial/plant metabolites, and water). Chromatographic techniques provide the most reliable data due to their precision and accuracy of analysis; therefore, they have also been recommended for use in evaluating alternative rapid techniques. Analytical methods should be appropriate and efficient for each matrix array, i.e., each modification introduced must be in accordance with validation criteria and the specific requests of regulatory organization.

Incomplete extraction and matrix effects of crude extract in the cleaning step can lead to a sub-estimation of real concentrations in analysis; thus, a minimum preparation is advantageous. The multi-toxin methods for HPLC and sequential mass spectrometry (LC-MS/MS) provides a high selectivity, lower limits of quantification and detection, the possibility of generating structural information of the analyte with minimal sample treatment, and the reduction of errors associated with pre-and post-column derivatization. Analyses by LC-MS/MS has gained much interest in analytics [14, 15].

Although there have been other advances in analytics, HPLC coupled with fluorescence and ultraviolet detectors remain the main detection method in Brazil [16, 17, 18]. In addition, the unavoidable occurrence of mycotoxins has obliged several countries to adopt regulatory guidelines, and maximum tolerated levels vary widely among countries [19].

Current regulations are increasingly based on international organizations, such as the FAO/WHO Joint Expert Committee on Food Additives of the United Nations (JECFA), and the European Commission. Strict guidelines on mycotoxins have been imposed by importing countries, demanding a rigorous and continuous monitoring of the food chain. The prevailing guidelines for mycotoxins require different protocols of extraction and analysis, and foods for infants and young children with more restrictive limits increases the number of analyses [20]. Such diversity in extraction procedures results in costly work.

Safe raw materials should be tracked by reliable analytical methods, and rapid methods are useful tools, especially in food-producing countries. Immunoassays based on ic-ELISA with highly specific monoclonal antibodies (MAb) against ochratoxin (OTA), fumonisin (FB), aflatoxin (AF), deoxynivalenol (DON), zearalenone (ZEA) and microcystin (MC) have been developed, previously tested for cross-reactivity with each analogue group, and correlated with HPLC as the primary method (Table 1, 2 and 3). A careful evaluation of ic-ELISA was conducted in the analysis of natural toxins in the food chain targeted to field/storage stage, beginning with the monitoring of fumonisins in corn [21]. The successful rapid technique motivated to use of this analysis for OTA in coffee and wine [22, 23], aflatoxin [24, 25], DON [26, 27, 28] and ZEA [29].

Hybridoma Cell line	Toxin	Cross reactivity (%)*	Hybridoma Cell line	Toxin	Cross reactivity (%)*
AF.2[30]	AFB_1	100	ZEN.2[34]	ZEA	100
	AFB_2	133		α-Zearalenol	60
	AFG_1	13.4		β-Zearalenol	5.7
	AFG_2	14.7		α-Zearalanol	7.1
	AFM_1	0.9		β-Zearalanol	0.9
DON.3[31]	DON	100	M8H5[35]	MCLR	100
	15-acetil DON	333		MCRR	106
	NIV	5		MCYR	44
	4-acetil NIV; Toxin T-2 tetraol	1.2		MCLA	26
	Others *	<0.5		3-desmethyl MCLR	51
				7-desmethyl MCLR	48
OTA.1 / OTA.7[32]	OTA	100 / 100		MCLR GSH conjugate	47
	OTC	63.1 / 79.4		MCLR methyl ester	30
	(4R)-4-HydroxyOTA	1.19 / 1.24		Nodularin	20
	OTB	0.63 / 1.07		6 (Z)-Adda -MCLR and - MCRR	<7
FB 1-2[33]	FB_1	100	MC. 5-3/ 8-3 / 2[36]	MCLR	100
	FB_2	224		MCRR	146 / 113 / 60
	FB_3	72		MCYR	88 / 65 / 113

AFB_1: Aflatoxin B_1; AFB_2: Aflatoxin B_2; AFG_1: Aflatoxin G_1; AFG_2: Aflatoxin G_2; AFM_1: Aflatoxin M_1; DON: Deoxynivalenol; NIV: Nivalenol; ZEA: Zearalenone; OTA: Ochratoxin A; OTC: Ochratoxin C; OTB: Ochratoxin B; FB_1: Fumonisin B_1; FB_2: Fumonisin B_2; FB_3: Fumonisin B_3; MCLR: Microcystin-LR; MCRR: Microcystin-RR; MCYR: Microcystin-YR; and MCLA: Microcystin-LA.

* Percentage of relative cross-reactivity was calculated as the amount of toxin required for 50 % binding inhibition/amount of other toxins requiring 50 % binding inhibition × 100.

** Others: 3-acetyl DON, 3,4-diacetyl NIV, tetraacetyl NIV, Toxin T-2, Toxin T-2 acetyl, and diacetoxyscirpenol.

[30] Kawamura et al., 1988; [31] Kawamura, 2005; [32] Kawamura et al., 1989; [33] Iijima et al., 1996; [34] Kawamura e Emoto, 2006; [35] Nagata et al., 1995; [36] Tabuchi et al., 2015.

Table 1. Cross-reactivity of monoclonal antibodies (anti-mycotoxins & microcystins) applied in a monitoring study in Brazil.

Table 1 shows the cross-reactivity of MAb (anti-mycotoxins & microcystins). It confirmed the high specificity of selected hybridomas, which were adequate for application in rapid surveys. Cross-reaction in immunoassays would be expected due to the biosynthesis of natural toxins in a sequential cluster of closely related structural substances. Nevertheless, the cross-reactivity within analogues can be advantageous in screening surveys of natural toxins compared with strongly specific individual analogue detection by HPLC.

Table 2 shows how ic-ELISA became established as reliable rapid technique to analyse mycotoxins and microcystins. Such local set-ups can allow safe supervision in one of the major food producing regions in Brazil, which was made possible due to joint research involving cell culture technologies, adaptation and proliferation of MAb producing hybridomas, and the development of immunoassays *in loco*. The standardized immunoassay was obtained through enhancing its sensitivity and adjusting to local conditions for the reagents, dilutions in ic-ELISA steps (upgrading crude extract preparation, antigen-protein conjugates for microplate coating, and dilutions of both the first and second antibody), and the analogue group for detection. The safety of the food, derived products, and water for analysis were amplified by awareness. These assays should be conducted for local consumption safety, the balance of agribusiness and exportation demand and importation independence.

Item	Hybridoma	Toxins	Reagents: ic-ELISA steps			LOD / LOQ ($\mu g\ kg^{-1}$)	ELISA/ HPLC (r)	Cereal & products
	Cell line	Mycotoxin/ Cyanotoxin	Coating	First MAb	Second Ab: IgG-enzyme			
1			DON-HG-OVA 2 $\mu g\ mL^{-1}$	1200 $\mu g\ mL^{-1}$	1:2000	177.1 / -	0.93	Wheat grain[26]; wheat flour[27]
2	DON.3	DON	DON-HS-OVA 2 $\mu g\ mL^{-1}$	19.2 $\mu g\ mL^{-1}$	1:1000	113.5 / 445.3		Wheat grain[28]
3			DON-HS-OVA 2 $\mu g\ mL^{-1}$	10.9 $\mu g\ mL^{-1}$	1:2000	159.3 / 370	-	Biscuit[a]
4	ZEN.2	ZEA	ZEN-OVA 2.5 $\mu g\ mL^{-1}$	10.3 $\mu g\ mL^{-1}$	1:2000	33.7 / 87	-	Wheat grain[a]
5			ZEN-OVA 2.5 $\mu g\ mL^{-1}$	10.3 $\mu g\ mL^{-1}$	1:2000	9.7 / 23.7	-	Biscuit[a]
6	FB 1-2	FB_1	FB1- OVA 0.77 $\mu g\ mL^{-1}$	1:50	1:5000	93 / -	0.94	
7	AF.2	AF	AFB_1-BSA 0.25 $\mu g\ mL^{-1}$	0.094 $\mu g\ mL^{-1}$	1:2000	2.0 / 4.6	-	Corn grain[a,21]
8	DON.3	DON	DON-HS-OVA 2 $\mu g\ mL^{-1}$	10.9 $\mu g\ mL^{-1}$	1:2000	302.8 / 589.3	-	
9	ZEN.2	ZEA	ZEN-OVA	10.3 $\mu g\ mL^{-1}$	1:2000	51.7 / 93.2	0.91	

Item	Hybridoma	Toxins	Reagents: ic-ELISA steps			LOD / LOQ ($\mu g\ kg^{-1}$)	ELISA/ HPLC (r)	Cereal & products
	Cell line	Mycotoxin/ Cyanotoxin	Coating	First MAb	Second Ab: IgG-enzyme			
			2.5μg mL^{-1}					
10	AF.2	AF	AFB_1-BSA 0.25 μg mL^{-1}	0.094 μg mL^{-1}	1:2000	1.25 / 1.43	0.97	Broiler feed[24]
11			AFB_1-BSA 0.25 μg mL^{-1}	0.094 μg mL^{-1}	1:2000	1.41 / 1.75	0.98	Laying hen feed[25]
12						0.17 / 0.32		Red wine[a]
13	OTA.1	OTA	OTA-BSA 0.077 μg mL^{-1}	0.043 μg mL^{-1}	1:1000	0.14 / 0.23	0.97	White wine[a]
14						0.17 / 0.32		Table wine[a]
15	OTA.7	OTA	OTA-BSA 4.76 μg mL^{-1}	1:2000	1:1000	3.75/ -	0.98	Green cofee[22]
16	M8H5	MCLR	MCLR-BSA 1:20000	1:20000	1:5000	- / 0.05	-	Fresh Water[37]

ic-ELISA: Indirect competitive enzyme linked immunosorbent assay.

DON: Deoxynivalenol; ZEA: Zearalenone; FB_1: Fumonisin B_1; AF: Aflatoxin; OTA: Ochratoxin A; MCLR: Microcystin-LR; DON-HG-OVA: Deoxynivalenol-hemiglutarate-ovalbumin; DON-HS-OVA: Deoxynivalenol-hemisuccinate-ovalbumin; ZEN-OVA: Zearalenone-ovalbumin; AFB_1-BSA: Aflatoxin B_1- Bovine Serum Albumin; OTA-BSA: ochratoxin A- Bovine Serum Albumin; and MCLR-BSA: Microcystin-LR - Bovine Serum Albumin.

[26]Santos et al. 2011; [27] Santos et al. 2013; [28] Souza et al. 2014; [21] Ono et al. 2001; [24] Rossi et al. 2013a; [25] Rossi et al. 2013b; [22] Fujii et al, 2006; [37] Kamogae et al. 2006; [a]Data not published.

Table 2. Development of ic-ELISA: standardized immunoassay for mycotoxins and microcystins analysis.

The optimized ic-ELISA showed a correlation coefficient of >0.9 with HPLC (Table 2). The result obtained with anti-OTA MAb produced by hybridoma OTA.1 was adequate to analyse wine using 1:10,000 anti-OTA MAb and 1:30,000 OTA-BSA. However, the matrix interference in the OTA analysis in wine by ic-ELISA should be considered. In analysing 60 wine samples, only one was OTA positive by HPLC (0.12 ± 0.01 ng mL^{-1}), whereas 11 false-positives were observed by ic-ELISA (range from 0.32 ± 0.02 to 0.47 ± 0.14 ng mL^{-1}). False-positive data in red wine may be attributed to the interference of anthocyanins and other pigments on OTA-binding to the antibody [38, 39]. The influence of matrix interference in OTA detection by ic-ELISA could be explained using a principal component analysis through the relationship of higher *trans*-resveratrol and OTA levels in the positive samples (Figure 2). In contrast, the addition of condensed tannins can inhibit the binding activity of antibodies in ELISA [40].

The undesired matrix effect and be minimized by diluting the crude extract prior to ic-ELISA; a 1:100 dilution of coffee extract minimized the matrix effect on OTA detection, regardless of the maturity stage [22]. Additionally, a dilution factor of 1:80 minimized the matrix effect when anti-DON MAb produced by Hybridoma DON.3 was used in ic-ELISA for wheat grain.

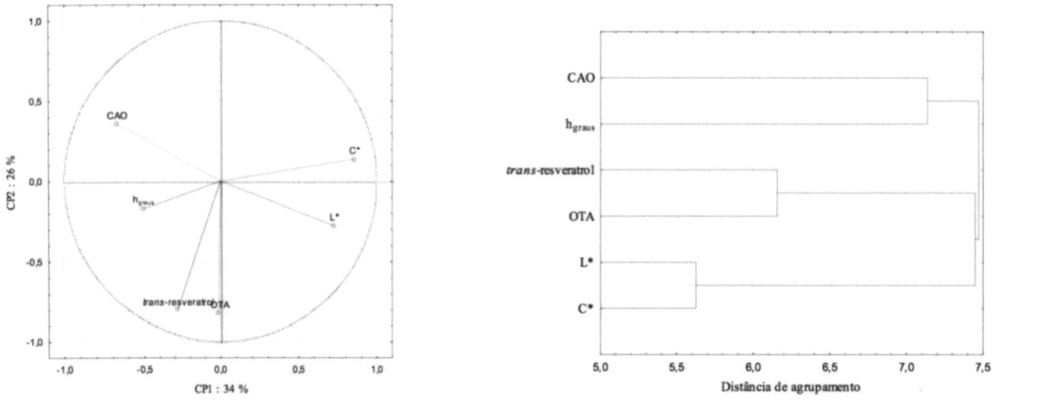

(a) Projection of variables. **(b)** Cluster analysis of parameters

Figure 2. Principal Component Analysis in the evaluation of wine samples, Paraná State: OTA (ochratoxin A), CAO (antioxidant capacity), *trans*-resveratrol and chromatic (h_{graus}, L*, C*). a) graphical representation in two dimensions (Principal Component 1- CP1 and Principal Component 2- CP2) and b) cluster analysis of parameters.

Table 3 shows the monitoring of natural toxins (mycotoxins & microcystins) by ic-ELISA developed for different food specimens, as well as in the freshwater since the 1990s. Corn, coffee, wheat, grain-derived products, wine, broiler and laying hen feeds, and fresh water in agricultural regions were analysed (Table 3). The table also shows some maximum limits established by Brazilian guidelines, the European Commission, and the World Health Organization.

Item	SAMPLING / MONITORING				Ic-ELISA		
	Cereal & products, Feed, Wine[41, 42]		Locality	Crop Year	Toxins Mycotoxin[a]	+/total (n)	Mean (μg kg-1)
1	Wheat	Grain	North-West, North-East, South-West /RS	2006 – 2008		15 / 15	2918.1
			North, Central, South-West /PR	2006 - 2008		7 / 23	1578.6
			North/PR	2009	DON	36 / 50	2379.4
		Flour	North/ PR	2009		21 / 23	2455.9
2		Grain	Central-South /PR	2010 - 2011		84 / 84	1879.3
			North /PR			159 / 160	848.9
3		Biscuits	North /PR (Retail Market)	2013		29 / 56	742.4
4		Grain	North, Central-South	2010 - 2011	ZEA	11 / 125	161.4

	SAMPLING / MONITORING				Ic-ELISA		
Item	**Cereal & products, Feed, Wine[41, 42]**		**Locality**	**Crop Year**	**Toxins Mycotoxin[a]**	**+/total (n)**	**Mean (μg kg-1)**
5		Biscuits	North (Retail Market)	2013		17 / 56	56.1
6	Corn	Grain	Central-South, Central-West, North / PR	1995 - 1996	FB_1	147 / 150	5610.0
7					AF	12 / 75	8.1
8			Central-South / PR	2010 - 2012	DON	6 / 75	2142.3
9					ZEA	36 / 75	522.3
10	Feed[44]	Broiler	North / PR	2010	AF	114 / 158	2.2 -6.4
11		Laying hen	North / PR	2010	AF	66 / 95	9.61
12	Wine	Red	North / PR (Retail Market)	2006 - 2011		8 / 47	0.42
13		White		2006 - 2009	OTA	4 / 23	0.68
14		Table wine	South-West, West	2010 - 2011		10 / 34	0.37
15	Coffee	Green coffee	North / PR	2003	OTA	15/68	5.28
					Cyanotoxin	+/total (n)	Mean (μg L-1)
16	Fresh Water[43]	Tibagi River	North / PR	1999 - 2000		13 / 24	0.28
		Itaipu Lake	West /PR		MCLR	22 / 24	18.35
*		Tibagi River	North /PR	2014		1 / 6	0.67
		Itaipu Lake	West / PR			3 / 6	0.65

ic-ELISA: Indirect competitive enzyme linked immunosorbent assay.

* These analysis were carried out using a commercial Kit (Beacon Microcystin Plate Kit, USA).

PR: Paraná State; RS: Rio Grande do Sul State.

[41] The Brazilian Health Surveillance Agency (ANVISA, 2011) established the maximum levels of DON in flour and biscuits (1,750.0 μg kg^{-1}), ZEA for biscuits (200.0 μg kg^{-1}), AF for corn (20.0 μg kg^{-1}), and OTA for wine (2.0 μg kg^{-1}). The deadlines were established in RDC 07/2011 and will be extended to January 1, 2017 for DON in wheat and corn (3,000.0 μg kg^{-1}), ZEA in wheat and corn (400.0 μg kg^{-1}), and $FB_1 + FB_2$ in corn (5,000.0 μg kg^{-1}) [42].

[43] The World Health Organization (WHO, 1998) established maximum levels of 1 μg of MCLR L^{-1} for drinking water and a Tolerable Daily Intake of 0.04 μg of MCLR kg^{-1} body weight.

[c] Mycotoxin / Cyanotoxin: DON, Deoxynivalenol; ZEA, Zearalenone; FB_1, Fumonisin B_1; AF, Aflatoxin; OTA, Ochratoxin A; and MCLR, Microcystin-LR.

[44] European Commission (2003). The maximum limit allowed by the European Commission is 0.02 mg aflatoxin B_1 kg^{-1}.

Table 3. Monitoring of natural toxins (mycotoxins & microcystins) by the rapid ic-ELISA method.

The application of ic-ELISA to monitoring freshly harvested corn from Paraná State (1991 to 2004 crops) indicated the widespread occurrence of fumonisins but a low occurrence of aflatoxins. In a recent study conducted in Paraná State, 74 corn samples were contaminated with an average of 1,840 μg of fumonisin kg^{-1}, 36 of poultry feeds with 239 μg of fumonisin kg^{-1}, and 9 corn factory residues with 23,676 μg of fumonisin kg^{-1}, whereas the aflatoxin and trichothecene levels were approximately at the LOD values [45]. Ic-ELISAs, using monoclonal mAb produced by hybridoma cells (AF.2, ZEN.2 and DON.3), were developed and optimized for AFs, ZEA and DON detection (Table 2). In corn samples from an experimental farm in central-southern Paraná State, 12 samples were found to be positive for AF (mean of 8.1 μg kg^{-1}), 36 samples for ZEA (mean of 522.3 μg kg^{-1}) and 6 samples for DON (mean of 2142.3 μg kg^{-1}) (Table 3).

An emphasis was placed on DON evaluations by ic-ELISA in wheat from 2006 to 2011 (Table 3). Paraná and Rio Grande do Sul States in southern Brazil produce 90 % of the national wheat [1]. This country depends on the importation of 5 to 6 million ton per year to provide for an annual domestic consumption of approx. 11 million ton, mainly used in bakery (55 %), pasta (17 %) and biscuit (13 %) processing [2, 46, 47]. Brazil is the world's second-largest biscuit producer, but the current low exportation (54,083 tons) results in nearly all production earmarked for domestic consumption, despite its ranking [48, 49]. In the wheat samples from experimental farms of north and central-southern of Parana State analysed by ic-ELISA, DON was detected in almost all of samples (243 positive samples of 244) and ZEA was detected in 10 of 125 samples (Table 3). In, wheat-based biscuits acquired from a local retail market in Londrina, Paraná State (56 samples) DON was detected in 29 samples (mean of 742.4 μg kg^{-1}) and ZEA in 17 samples (mean of 56.1 μg kg^{-1}) (Table 3). A study [50] analysed 23 cracker biscuit samples produced in Southern Brazil and group A trichothecene was non-detectable, but 18 samples were contaminated with DON (378 – 5295 μg kg^{-1}), with 22 % of the samples at level over the Brazilian guideline limit (1,750 μg kg^{-1}). When zearalenone was analysed in corn-based products (51 samples of popcorn and 50 corn grits) and cracked wheat (n = 109) commercialized in 18 counties of the Paraná state, ZEA was non-detected in cracked wheat samples, but one cracked corn sample contaminated 64 μg of ZEA kg^{-1} [50]. Fusariotoxin monitoring in wheat should be conducted in both domestic production and in imported wheat, which represents 50 % of the category.

Due to the possible carry-over of mycotoxins to tissues, the degree of exposure of broiler chicken and laying hens to fumonisins and aflatoxins through naturally contaminated feeds has been assessed (Table 3). Occurrence of fumonisins and aflatoxins were evaluated in four feed types intended for broilers (n=158), collected from a poultry breeding farm in Northern Paraná State [24]. Fumonisins were detected in 94.9 % of the feed samples at mean levels ranging from 0.52 μg g^{-1} (finisher) to 0.68 μg g^{-1} (pre-starter and grower), and aflatoxins were detected in 72.1 % of the feed samples at mean levels ranging from 2.22 ng g^{-1} (pre-starter) to 6.41 ng g^{-1}(grower). The maximum estimated daily intake of FB_1 for broilers (0.057 mg/kg body weight/day) was below the Lowest Observed Adverse Effect Level (2 mg kg^{-1} body weight day^{-1}). Most of the aflatoxin positive samples (97 %) showed levels below the maximum limit allowed by the European Commission (0.02 mg aflatoxin B_1 kg^{-1}). To estimate the degree of exposure of laying hens to mycotoxins, a total of 95 mash feed samples were collected from January to December 2010 from the Experimental Farm at the University, Northern Paraná State, Brazil. Aflatoxins and fumonisins were detected in 69.7 % and 89.5 % of the feed (n=95) intended for laying hens at mean levels of 9.61 ng g^{-1} and 1.28 μg g^{-1}, respectively. The estimated daily intake of FB_1 for laying hens (0.038 mg kg^{-1} body weight day^{-1}) was below the Lowest Observed Adverse Effect Level (2 mg kg^{-1} body weight day^{-1}). Aflatoxin levels were below the maximum allowed limit by the European Commission in the majority of the positive samples (85.1 %), which indicated that some of the feed samples could have a negative effect on animal health and performance, but the risk would be very low.

Intensive agricultural activity has become an increasing concern due to the eutrophication of aquatic environments. Microcystins (MCs) were monitored in Itaipu Lake and Tibagi River in the north and west of Paraná State, respectively (1999 to 2000 and 2014). The reduction of microcystin levels in Itaipu Lake was likely a consequence of ecological programs encouraging the recovery of riparian forests, in addition to a change in planting management (Table 3). Such a reliable MAb-based rapid immunoassay has been a good choice for tracking mycotoxins and cyanotoxins and determining actions to be performed in a control strategy.

4. Monitoring strategies: The importance of reliable analysis as controlling guidance

Monitoring of the corn chain in northern Paraná showed that 81 % (n = 435, crop 2003) and 98.8 % (n = 435, crop 2004) of corn was safe for human consumption, in regard to fumonisin. The decreasing trend in fumonisin contamination, when compared to previous studies [8, 9, 12], could suggest a conscious monitoring procedure at the quality control level, in accordance with the strict guidelines imposed by importing countries.

The main approaches in corn phytosanitary control involve pesticides and agricultural practices with an emphasis on tillage and crop rotation. Efforts have been focused on novel fungicides for *Fusarium* sp. control to maximize grain yield [51, 52]; however, several studies have shown that fungicide application can increase mycotoxin levels [53, 54]. The recommended dose of fludioxonil + metalaxyl-M (2.5 + 1.0 %) was insufficient to inhibit *F. verticillioides* growth *in vitro*, but it increased FB_1 production from 3.5-fold to 12.5-fold (2.58 μg mL^{-1} when compared with 0.72 μg mL^{-1} in control), with an alteration in mycelial morphology [55, 56]. A scanning electron microscopy analysis showed that the fungicide caused the inhibition of hyphal growth and defects of hyphae, such as excessive septation, cell wall disruption, and withered hyphae, and extracellular material around the hyphae was rarely observed (Figure 3) [57].

Therefore, efforts to reduce mycotoxin levels should be focused on sustainable production. In uninterrupted planting in tropical regions, non-drastic management of cropping systems using culture rotation in no-tillage areas under different fertilizations emphasizing nitrogen rate, and low cost organic waste remain concerns in the protection of grain and soil conservation [58, 59, 60, 28].

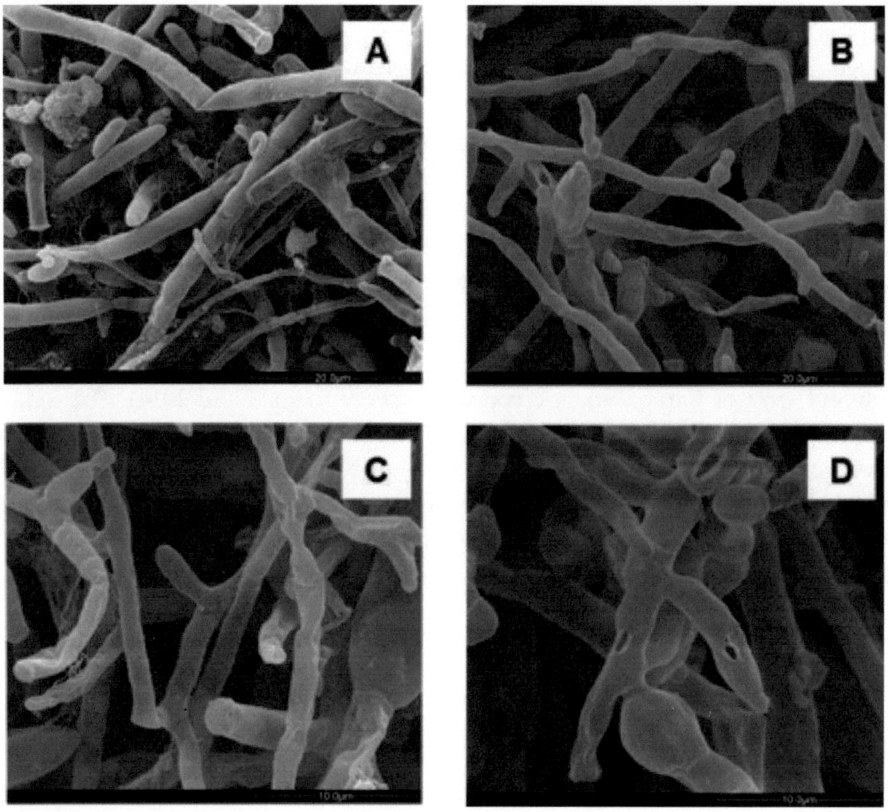

Figure 3. Electron micrographs of *F. verticillioides* 103 F mycelia cultured in defined liquid media in the absence (control) and presence (treatment) of fludioxonil + metalaxyl - M at the dose recommended by the manufacturer (1.5 μL mL^{-1}) showing the fibrillar extracellular material present in the control cultures (A and C) and withered hyphae and disruption of cell walls in the treatments (B and D).

The effect of conventional and no-tillage cropping systems in corn cultivated in summer following either oats or fallow in winter on natural fumonisin levels (2006 and 2007 growing seasons) has been assessed [60]. No-till corn following oats showed stronger fumonisin contamination patterns than the other treatments (2006 season, $P<0.05$). Although no-till could be advantageous from a soil conservation standpoint, it may enhance fumonisin contamination in the tropics, contrasting another report [61] that there was no significant difference between conventional and no-till in fumonisin in monoculture corn in Northern Italy. When the nitrogen fertilizer rates (0 to 90.0 kg ha^{-1} N) on fumonisin contamination was evaluated, higher fumonisin levels were detected in plots with lower N (≤ 22.5 kg ha^{-1}) than ≥ 45.0 kg ha^{-1} N, indicating a negative correlation between fumonisin and N rates [60]. Both N stress due to deficiency and excessive rates can increase the FB_1 level in corn [62].

The nitrogen-fixing potential of *Azospirillum* sp. in the rhizosphere can increase yields, reduce costs and improve the nutritional quality of corn kernels. An experiment was conducted matching the inoculums of the *Azospirillum brasilense* Ab-V5 and Ab-V6 strains in corn seeds with N doses in Northern and Central-Southern Paraná in the 2010/2011 and 2011/2012 seasons [63]. Although the seed inoculation associated with N doses showed non-significant effects on fungal count, the inoculum treated plots showed lower fumonisin levels ($p < 0.05$) than the non-treated plots, indicating favourable trend towards agricultural practices with inoculants. Fumonisins were detected in 90 % of samples in 2010/2011 (mean, 0.62 $\mu g\ g^{-1}$) and 97.5 % in 2011/2012 (mean, 4.34 $\mu g\ g^{-1}$) in the northern region, whereas its occurrence was 45 % in 2010/2011 (mean, 0.14 $\mu g\ g^{-1}$) and 100 % in 2011/2012 (mean, 2.67 $\mu g\ g^{-1}$) in central-southern Paraná.

The use of landfill leachate in agricultural soils as fertilizers has been suggested as an alternative for the disposal of this effluent; however, heavy metals may be a limiting factor [63]. The application of increasing doses of leachate (0 to 130.8 $m^3\ ha^{-1}$) increased the yield, protein content, lipid and ash in corn grain, but no effect was observed on fumonisin reduction, which occurred in all samples, with 31.2 of samples with levels over the maximum tolerable limit in Brazilian guidelines (5.0 $\mu g\ g^{-1}$). An increasing trend in lead content was also observed in the 2009/2010 seasons, and in sodium (2011/2012 seasons) when the leachate rate was increased [63].

The management of plant density (60 to 105 thousand plants/ha) with N doses (0 to 240 kg ha^{-1}) showed no effect on corn fungal count, but there was an increasing trend in fumonisin levels when plant density was increased. Total fumonisins (FB_1 + FB_2) were detected in corn grain at levels ranging from non-detectable to 7.80 $\mu g\ g^{-1}$ (mean, 1.50 $\mu g\ g^{-1}$) in the 2009/2010 season, while it was non-detectable to 23.36 $\mu g\ g^{-1}$ (mean, 1.72 $\mu g\ g^{-1}$) in the 2010/2011 season [63].

Efforts also should be focused on the safety and quality of the wheat chain, one of major universal components in food. Although 90 % of the national crop is centred in southern Brazil, domestic consumption still depends on importation [1, 2]. Table 3 shows that natural contamination of DON in wheat was non-equally distributed among different crops and was dependent on local and climatic conditions (the impact of agricultural management practices was evaluated in 2010 and 2011 seasons). Environmental conditions can shift the metabolic route of *Fusarium graminearum*, changing the fusariotoxin profile in grains, ex., leading to an increase of acetylated analogues of DON [64, 65]. Although such acetylated trichothecenes are considered less toxic than DON, a rapid deacetylation can take place in the digestive tract of mammals, turning into it to DON [66]. DON of the group B has been regarded as a unique trichothecene in wheat products under current Brazilian guidelines [41], although studies have indicated that group B trichothecenes, such as nivalenol (NIV) and acetylated analogues (3-acetyl-DON and 15-acetyl-DON), should be included [17].

In addition, lactic bacteria of the *Lactobacillus plantarum* group have shown versatile profiles concerning genes that can code for special functions such as biodegradation, absorption, and adherence to different surfaces in the vast natural microbiota in food niches [69, 68]. Lactic acid bacteria strains can be isolated from multiple wheat sources (grains, germ, bran, and flour) and have been tested against *F. graminearum* strain IAPAR 2218 (Figure 4). All tested strains demonstrated some DON-reducing potential, and the non-viable autoclaved *L. plantarum* cells (71.19 %) showed a higher effectiveness in DON reduction than viable cells (16.41 %). The probable mechanism of reduction would be the adsorption by cell walls [69, 70]. Lactic acid bacteria can also degrade a range of low molecular weight compounds, carrier compound families, and influx and efflux facilitators and mycotoxin degrading enzymes have been detected [71].

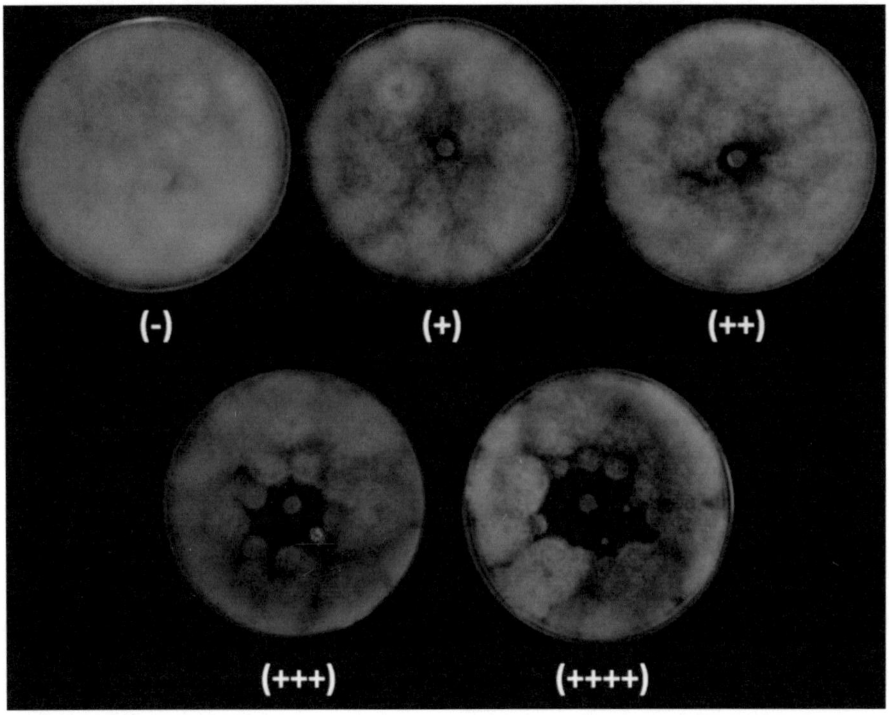

Figure 4. Effect of lactic acid bacteria isolated from different wheat sources (grains, germ, bran, flour) against *F. graminearum* strain IAPAR 2218. Inhibition Halo scale: (-): no inhibition; (+):1 to 5 mm; (++): 6 to 10 mm; (+++): 11 to 15 mm; and (++++): >15 mm.

Another use of naturally occurring microorganisms in the biocontrol/biodegradation of undesired natural toxins has been assessed for the reduction of cyanobacteria in drinking water. The potential of microcystin (MC) biodegradation has been tested in the following microorganisms: *Sphingosinicella microcystinivorans* (B9) isolated of the Lago Tsukui, Kanagawa-Japan; water kefir (mixture of lactic and acetic bacteria and yeast) (P4); *L. acidophilus* La-5 (P5); and yeast isolated of sugarcane (L5). The strain B9 degraded 99 % of MCs, while the strains P4, P5 and L5 degraded 44, 43 and 54 % of total MCs, respectively, after 96 h (Figure 5).

Strain B9 (*S. microcystinivorans*) showed the highest MCs degradation capacity and has been evaluated for its anti-cyanobacterial activity against 5 cyanobacteria strains, *Microcystis* sp. (C1), *Microcystis* sp. (C2), *Anabaena ucrainica* (C3), *Phormidium tenue* (C4), and *Synechocystis* (C5). After 96 h, the inhibition percentages (cellular counts) against cyanobacteria strains ranged from 41.4 to 79.3 %, while the inhibition percentages (concerning chlorophyll-a) ranged from 34.4 to 68.9 %.

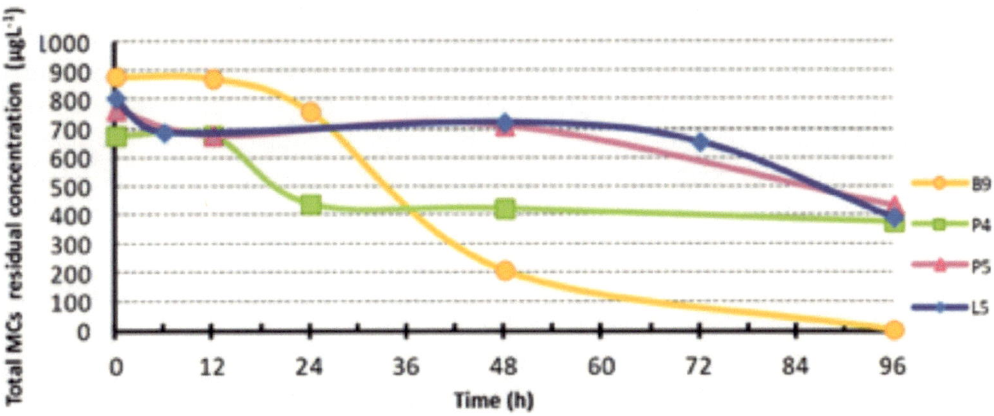

Figure 5. Biodegradation of microcystin by *Sphingosinicella microcystinivorans* (strain B9), water kefir (P4), *Lactobacillus acidophilus* La-5 (P5) and yeast (L5).

In summary, adequate agricultural practices based on crop rotation, fertilization, soil biodiversity, resistant crops, and post-harvest management could reduce mycotoxin contamination in field. Further long-term strategies encouraging no-tillage cultivation, and the maintenance of riparian forests in extensive agricultural land would be goals to maintain water quality; and the sustainable production of nutrient-rich high-quality products would still be possible.

5. Novel emerging tools for quality & safety: non-destructive technology in the food chain

Traditional analytical techniques for food and feed quality inspection and compositional assessment are typically invasive and time-consuming, requiring extensive sample preparation, thus being unsuitable for applications in the highly demanding, fast-paced food processing segment. Recently, novel techniques have been investigated for fast, reliable and chemical-free food quality assessments. Near-infrared (NIR) hyperspectral imaging has emerged as an efficient and advanced tool, combining both computer vision techniques and NIR spectroscopy, which can be used for continuous monitoring, process control and quality assessments of agricultural products, food and feed materials. Because most food quality features are related either to the external appearance of the product or its chemical composition, either computer vision or NIR spectroscopy alone is adequate for monitoring organic samples in a fast, reliable manner. However, such techniques are still strongly dependent on other reference methods. Prediction of physical characteristics and chemical composition using NIR spectroscopy and/or computer vision methods has been reported on meat (chicken, pork, beef, lamb), cereals and grains (corn, wheat, soy, rye, coffee, cocoa), and fruits and vegetables (apple, citrus, berries). More specifically, there has been major interest in this technique in quality control, food safety and security, i.e., detection and prediction of contamination in agricultural products.

A study [72] compared NIR calibration methods for predicting protein, oil and starch contents in both whole and ground maize samples in the spectral range of 1100–2500 nm for reflectance and 680–1235 nm in transmittance modes. While the best models were obtained for the reflectance spectra of the ground samples, it was suggested that the transmittance mode for whole grains might be more useful due to its greater speed of analysis. Another study [73] developed a rapid single kernel NIR sorting instrument for maize and soybean. Prediction models for moisture of both seed types, and protein contents for soybeans were developed utilizing a spectrometric range from 906 to 1683 nm.

NIR reflectance and transmittance technologies have been investigated for contamination assessments of a range of cereal grain physical quality and chemical traits, and detecting and predicting levels of mycotoxins. Numerous applications have been developed, and cover almost all cereals in the globally important food grains, i.e., corn, wheat, rice and barley. An additional application has been to demonstrate the value in sorting grains infected with fungus or mycotoxins, such as deoxynivalenol, fumonisins and aflatoxins [74].

A shortwave infrared (SWIR) hyperspectral imaging system in the wavelength range between 1000 and 2500 nm was used to assess the potential AFB_1 contaminants on the surfaces of healthy corn kernels. Key wavelengths that can indicate AFB_1 and are used to differentiate levels of AFB_1 were identified. A minimum classification accuracy of 88 % was achieved for the validation set and verification set, indicating that hyperspectral imaging technology could be used to detect AFB_1 at levels as low as 10 ng/g, when applied directly on the corn surface [75]. Another study assessed the applicability of NIR for the rapid identification of mycotoxigenic fungi and their toxic metabolites produced in naturally and artificially contaminated products. Two hundred and eighty corn samples were collected in north-central Italy and analysed for fungal infection, ergosterol, and FB_1 content. The results indicated that NIR could predict the incidence of kernels infected by *F. verticillioides* and also the quantity of ergosterol and fumonisin B_1 in the meal. The best predictive ability for the percentage of global fungal infection and *F. verticillioides* was obtained using a calibration model utilizing corn kernels (r 2 0.75 and SECV 7.43) and maize meals (r 2 0.79 and SECV) 10.95), respectively [76].

A recent study on the quality assessments of meat products [77] reported the application of NIR reflectance as a potential method to predict quality attributes of chicken breast (*Pectoralis major*). Spectra in the wavelengths between 400 and 2500 nm were analysed, presenting clear differences between different quality grades of chicken (Figure 6). PCA performed on the NIR dataset revealed the influence of muscle reflectance (L*) influencing the spectra. PCA was not successful to completely discriminate between pale, soft and exudative (PSE) and pale-only muscles. High-quality PLSR were obtained for L* and pH models predicted individually (R2CV of 0.91 and 0.81, and SECV of 1.99 and 0.07, respectively). Sample mincing and different spectra pre-treatments were not necessary to maximize the predictive performance of the models. The results suggest that NIR spectroscopy may represent a useful tool for the quality assessment of chicken meat.

The contamination of meat products has also been investigated. NIR transflectance and Fourier transform-infrared (FT-IR) attenuated total reflectance spectra of intact chicken breast muscle were collected and investigated for their potential use in the rapid, non-destructive detection

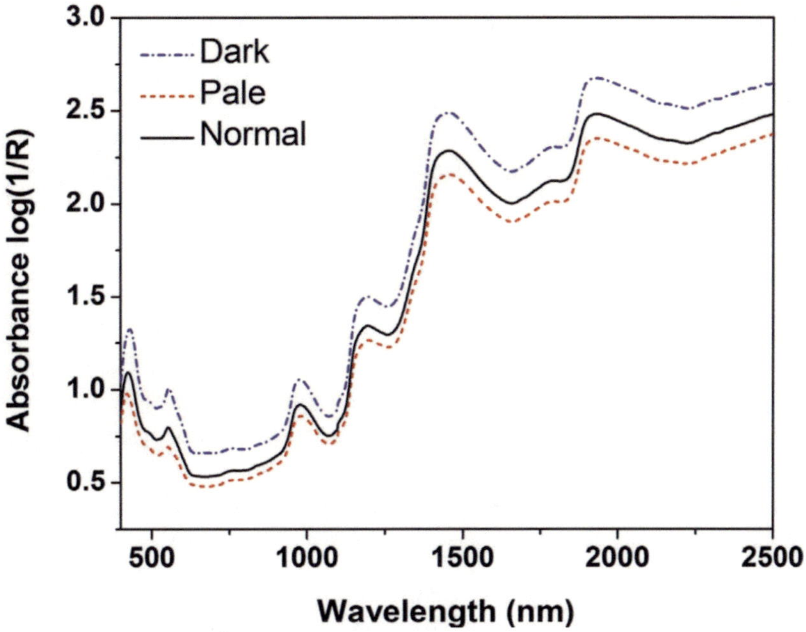

Figure 6. Average spectra for dark, normal and pale chicken samples (see [76] - "Reprinted from Food Chemistry, 168, Douglas Fernandes Barbin, Cintia Midori Kaminishikawahara, Adriana Lourenco Soares, Ivone Yurika Mizubuti, Moises Grespan, Massami Shimokomaki, Elisa Yoko Hirooka, Prediction of chicken quality attributes by near infrared spectroscopy, 554-560, 2015, with permission from Elsevier").

of spoilage. PCA and PLS2-DA regression correctly classified 8 and 14 day samples (TVC days 8 and 14= 9.61 and 10.37 log10 CFU g^{-1}) with several correlations that highlight the effect of proteolysis influencing the spectra. These correlations indicate that an increase in free amino acids and peptides could be the main factor in the discrimination of intact chicken breast muscle.

These studies have demonstrated that NIR methodology can be applied to monitor bacterial and fungal contamination in postharvest grains and fresh meats and to distinguish contaminated from clean batches to avoid cross-contamination with other materials during storage. However, there is still a demand for the development of cost-effective technologies for high-speed sorting. In the area of food safety, it is important to create robust prediction models based on reference methods by including a wide range of samples from different regions. For instance, it is well known that a major drawback of this technology is the application of ready-to-use prediction models from one country into samples from another region of the world. Prediction models are usually built in developed countries, but they are not useful for samples originating in developing countries, mostly due to inherent differences in sample composition, cultivation methods, climate and soil characteristics, etc. Once researchers overcome these obstacles, this technology will benefit farmers, the industry and consumers if it enables contaminated grain and other food samples to be identified and removed from the food chain.

6. Trends in globalized agribusiness

Globalization demands quality and competitiveness throughout the food chain, as well as safe raw materials without deterioration. Agribusiness exports in Brazil reached US$ 5.64 billion in January 2015, according to the foreign trade statistics system of Brazilian agribusiness [78]. The five main exporter states in Brazil were São Paulo, Paraná, Mato Grosso, Minas Gerais and Rio Grande do Sul, whose participation represented approximately 69 % of total Brazilian exports, involving soy, sugar-alcohol complex, beef, chicken meat, soybean oil, cereal sales, and coffee.

Brazil is currently the third-leading country in chicken production, behind the USA and China, and the leading exporter of chicken meat [79]. The long-term trade projection estimates a production of up to 20.576 million tons by 2023, corresponding to an increase of 46.4 % between the years 2013 and 2023. In addition, the exportation of meat has been forecasted at 4.675 million tons in 2023 [80]. Further, poultry (broilers and turkeys) trade long-term projections of the USDA (2015) indicate that exportation could reach up to 4.982 million tons in 2024.

In this scenario of globalized agribusiness, corn stands out as the major component of animal feed production [4]. Figure 7 shows a scenario concerning corn purchased as a raw material in a potential importer country. The trend for importing of Brazilian grains continues to depend on the decrease or delay of the American harvest of agriproducts due to the unmatched infrastructural facilities and the storage, transport and port system established in the United States.

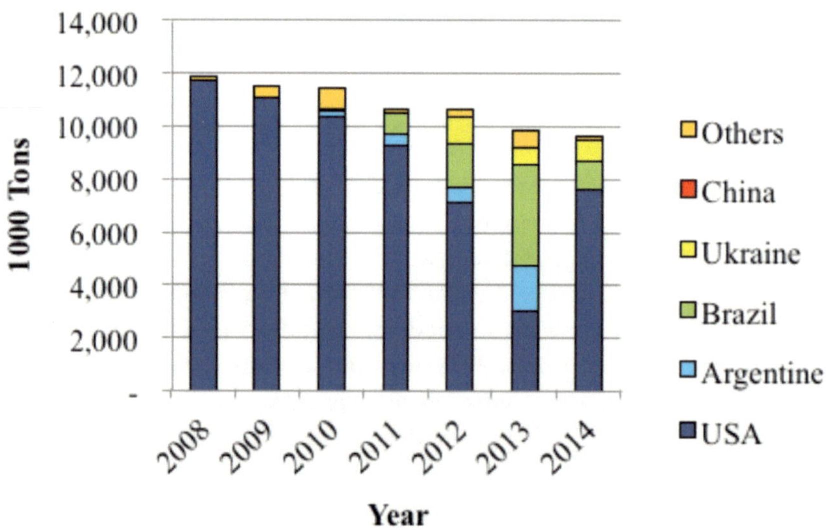

Figure 7. Corn importation in Japan, 2008 to 2014 [81].

Even with infrastructural problems, the bar chart shows that the decision to import Brazilian raw materials could unexpectedly increase and reached nearly five times the quota between the year 2012 and 2013.

The current effort on aggregation of value in agriproducts should be combined with continuous work on safety in highly productive regions, ex. broiler chickens for export, which depends on practical and reliable analytics facilitating quality and safety. Such an overall approach could result in promising healthy foods from potential producer regions in Brazil.

Acknowledgements

The authors thank the NANOBIO/CAPES Foundation (the Co-ordination for Formation of High Level Professionals) - Ministry of Education, the CNPq (the Brazilian Government Organization for grant aid and fellowship to Brazilian researchers) - Ministry of Science & Technology, the CNPq/Ministry of Agriculture, the Araucaria Foundation, and the Parana State Fund/SETI for the financial support through projects. The authors also thank the CNPq for the productivity fellowships granted to researchers, as well as the CAPES for fellowships at the levels of Postdoctorate, Doctorate and Initiation in Science and JICA (Japan International Cooperation Agency) for distinctive training support in Japan. A special thanks to Dr Yoshio Ueno (*in memoriam*) for introducing us to the advanced area of Microbial Toxins.

Author details

Elisa Y. Hirooka[1*], Elisabete Y.S. Ono[1], Cássia R. Takabayashi-Yamashita[1], Angélica T. Ishikawa[1], Douglas F. Barbin[2], Jaqueline G. Bordini[1], Melissa T. Hirozawa[1], Wagner E. Risso[1], Ana L.S.M. Felício[1], Artur K. Bagatin[3], Tatiana Á. Miguel[1], Thiago M. Souza[4], Yumi Tabuchi[3], Josemarque L. Rosa[1], Leonardo F. Maciel[1,5], Dani L. D. Silva[6], Renata P. Sobottka[7], André M. Prando[8], Sandra Garcia[1], Claudemir Zucareli[1], Emilia K. Kuroda[1], Elisabete H. Hashimoto[9], Diva de Souza Andrade[10], Martha Z. Miranda[11], Maria S. Miranda[4], Eliete S. Bispo[4], Adriana L. Soares[1], Massami Shimokomaki[12], Myrna Sabino[13], Eiko N. Itano[1], Tomoaki Tsutsumi[14], Satoshi Nagata[15], Yoshitsugu Sugiura[16], Ken-Ichi Harada[17] and Osamu Kawamura[3]

*Address all correspondence to: elisahirooka@hotmail.com

1 State University of Londrina, Londrina, Brazil

2 University of Campinas, Campinas, Brazil

3 Kagawa University, Miki, Japan

4 Oklahoma State University, Stillwater, EUA, USA

5 Federal University of Bahia, Salvador, Brazil

6 University Hospital, Londrina, Brazil

7 Agrária Agroindustrial Cooperative, Entre Rios–Guarapuava, Brazil

8 EMBRAPA CNPSo, Londrina, Brazil

9 Federal University of Technology-Paraná, Francisco Beltrão, Brazil

10 Agronomic Institute of Paraná, Brazil

11 EMBRAPA, Passo Fundo, Brazil

12 Federal University of Technology-Paraná, Londrina, Brazil

13 Adolfo Lutz Institute, São Paulo, Brazil

14 National Institute of Health Sciences, Tokyo, Japan

15 Cancer Biology Research Center, South Dakota, USA

16 Kobe Institute of Health, Kobe, Japan

17 Meijo University, Nagoya, Japan

References

[1] Companhia Nacional de Abastecimento (CONAB). 2013. Acompanhamento de safra brasileira: grãos, décimo segundo levantamento, setembro 2013. Available from: http://www.conab.gov.br/OlalaCMS/uploads/arquivos/13_10_16_14_32_01_ boletim_portugues_-_setembro_2013.pdf. [Accessed: 2013/11/04].

[2] Companhia Nacional de Abastecimento (CONAB). 2015. Acompanhamento da safra brasileira – grãos – safra 2014/2015, v.2, 1-117.

[3] Companhia Nacional de Abastecimento (CONAB). 2014. Acompanhamento da safra brasileira – grãos – levantamentos de safra – 2° levantamento - intenção de plantio – novembro 2014. Available from: http://www.conab.gov.br/OlalaCMS/uploads/ arquivos/. [Accessed: 2014/11/18].

[4] Associação Brasileira das Indústrias de Milho (ABIMILHO). 2011. O cereal que enriquece a alimentação humana. Available from: http://www.abimilho.com.br/ milho/cereal. [Accessed: 2014/01/14].

[5] Cawood ME, Gelderblom WCA, Vleggaar R, Behrend Y, Thiel PG, Marasas WFO. Isolation of the fumonisin mycotoxins: a quantitative approach. Journal of Agricultural and Food Chemistry. 1991;39:1958-1962. DOI: 10.1021/jf00011a014

[6] Diener UL, Cole RJ, Sanders TH, Payne GA, Lee LS, Klich MA. Epidemiology of aflatoxin formation by Aspergillus flavus. Annual Review of Phytopathology. 1987;25:249-270. DOI: 10.1146/annurev.py.25.090187.001341

[7] Sydenham EW, Marasas WFO, Shepard GS, Thiel PG, Hirooka EY. Fumonisin concentrations in Brazilian feeds associated with field outbreaks of confirmed and suspected animal mycotoxicoses. Journal of Agricultural and Food Chemistry. 1992;40:994-997. DOI: 10.1021/jf00018a016.

[8] Hirooka EY, Yamaguchi MM, Aoyama S, Sugiura Y, Ueno Y. The natural occurrence of fumonisins in Brqasilian corn kernels. Food Additives and Contaminants. 1996;13:7173 - 183. DOI: 10.1080/02652039609374396.

[9] Ono EYS, Ono MA, Funo FY, Medina AE, Oliveira TCRM, Kawamura O, Ueno Y, Hirooka EY. Evaluation of fumonisin-aflatoxin co-occurrence in Brazilian corn hybrids by ELISA. Food Additives and Contaminants. 2001;18:719-729. DOI: 10.1080/02652030118906.

[10] Silva M, Garcia GT, Vizoni E, Kawamura O, Hirooka EY, Ono EYS. Effect of time interval from harvesting to the pre-drying step on natural fumonisin contamination in freshly harvested corn from the State of Paraná, Brazil. Food Additives and Contaminants. 2008;25:642-649. DOI: 10.1080/02652030701618310.

[11] Ono EYS, Silva M, Hashimoto EH, Vizoni E, Kawamura O, Sugiura Y, Hirooka EY. Mycotoxicological quality evaluation of corn samples used by processing industries in the Northern region of Paraná State, Brazil. Food Additives and Contaminants. 2008;25:1392-1399. DOI: 10.1080/02652030802136204

[12] Ono EYS, Biazon GT, Silva M, Vizzoni E, Sugiura Y, Ueno Y, Hirooka EY. Fumonisins corn: correlatoon witu Fusarium sp. Count, damaged kernels, protein and lipid content. Brazilian Archives of Biology and Technology. 2006;49:63-71. DOI: http://dx.doi.org/10.1590/S1516-89132006000100008

[13] Moreno EC, Garcia GT, Ono MA, Vizoni E, Kawamura O, Hirooka EY, Ono EYS. Co-occurrence of mycotoxins in corn samples from the Northern region of Paraná State, Brazil. Food Chemistry. 2009;116:220-226. DOI: 10.1016/j.foodchem. 2009.02.037.

[14] Ren Y, Zhang Y, Shao S, Cai Z, Feng L, Wang Z. Simultaneous determination of multi-component mycotoxin residues in foods and feeds by ultra performance liquid chromatography tandem mass spectrometry. Journal of Chromatography A, 2007, 1143:48–64.

[15] Berthiller F, Sulyok M, Krska R, Schuhmacher R. Chromatographic methods for the simultaneous determination of mycotoxins and their conjugates in cereals. International Journal of Food Microbiology. 2007;119(1-2):33-7.

[16] Dors GC, Bierhals VS, Badiale-Furlong E. Parboiled rice: chemical composition and the occurrence of mycotoxin. Science and technology. 2011;31:172-177. DOI: http://dx.doi.org/10.1590/S0101-20612011000100025

[17] Del Ponte EM, Garda-Buddon J, Badiale-Furlong E. Deoxynivalenol and nivalenol in commercial wheat grain related to Fusarium head blight epidemics in southern Brazil. Food Chemistry. 2012;132:1087–1091. DOI: 10.1016/j.foodchem.2011.10.108.

[18] Baquião AC, Zorzete P, Reis TA, Assunção E, Vergueiro S, Correa B. Mycoflora and mycotoxins in field samples of Brazil nuts. Food Control. 2012;28:224-229. DOI: Mycoflora and mycotoxins in field samples of Brazil nuts.

[19] Food and Agriculture Organization of the United Nations (FAO). 2004. Worldwide regulations for mycotoxins in food and feed in 2003. Rome, Italy.

[20] Desmarcheliera A, Tessiota S, Bessairea T, Racaulta L, Fioreseb E, Urbanib A, Chanc WC, Cheng P, Mottieraa P. Combining the quick, easy, cheap, effective, rugged and safe approachand clean-up by immunoaffinity column for the analysis of 15mycotoxins by isotope dilution liquid chromatography tandem mass spectrometry. Journal of Chromatography A. 2014;1337:75-84. DOI: 10.1016/j.chroma.2014.02.025

[21] Ono EYS, Kawamura O, Ono MA, Ueno Y, Hirooka EY. A comparative study of indirect competitive ELISA and HPLC for fumonisin detection in corn of the State of Paraná, Brazil. Food and Agricultural Immunology. 2001;12:5-14. DOI: 10.1080/09540100099580.

[22] Fujii S, Ribeiro RMR, Scholz MBS, Ono EYS, Prete CEC, Itano EN, Ueno Y, Kawamura O, Hirooka EY. Reliable indirect competitive ELISA used for a survey of ochratoxin A in green coffee from the North of Paraná State, Brazil. Food Addittives and Contaminants. 2006;23:902-909. DOI: 10.1080/02652030600771509.

[23] Silva D. L. Anticorpo monoclonal de alta eficiência no desenvolvimento de coluna de imunoquímica aplicada- análise de ocratoxina em vinho [thesis]. Londrina: Universidade Estadual de Londrina; 2010.

[24] Rossi CN, Takabayashi CR, Ono MA, Bordini JG, Kawamura O,Vizoni E, Hirooka EY, Ono EYS. Assessment of exposure of broiler chicken in Brazil to mycotoxins through naturally contaminated feed. Food Security. 2013;5:541-550. DOI: 10.1007/s12571-013-0278-4.

[25] Rossi CN, Takabayashi CR, Ono MA, Bordini JG, Itano EN, Kawamura O, Pinheiro JW, Hirooka EY, Ono EYS. Exposure of laying hens to mycotoxins through naturally contaminated feed. World Mycotoxin Journal. 2013;6:199-207. DOI: 10.3920/WMJ2012.1511

[26] Santos JS, Takabayashi CR, Ono EYS, Itano EN, Mallmann CA, Kawamura O, Hirooka EY. Immunoassay based on monoclonal antibodies versus LC–MS: Deoxynivalenol in wheat and flour in Southern Brazil. Food Additives & Contaminants. 2011;28:1083–1090. DOI: 10.1080/19440049.2011.576442.

[27] Santos JS, Souza TM, Ono, EYS, Hashimoto EH, Bassoi MC, Miranda MZ, Itano EN, Kawamura O, Hirooka EY. Natural occurrence of deoxynivalenol in wheat from Paraná State, Brazil and estimated daily intake by wheat products. Food Chemistry. 2013;138:90-95. DOI: 10.1016/j.foodchem.2012.09.100.

[28] Souza TM, Prando AM, Takabayashi CR, Santos JS, Ishikawa AT, Felício ALDSM, Itano EN, Kawamura O, Zucareli C, Hirooka EY. Composição química e desoxinivalenol em trigo da região Centro-Sul do Paraná: adubação nitrogenada em cobertura associada com Azospirillum brasiliense. Semina Ciências Agrárias. 2014; 35:327-342. DOI: http://dx.doi.org/10.5433/1679-0359.2014v35n1p327.

[29] Takabayashi-Yamashita CR. Hybridoma generation and immunoassay employing monoclonal antibodies for natural toxins detection [thesis]. Londrina: State University of Londrina; 2013.

[30] Kawamura O, Nagayama S, Sato S, Ohtani K,Ueno I, Ueno Y. A monoclonal antibody-based enzyme-linked immunosorbent assay of aflatoxin B1 in peanut products. Mycotoxin Research. 1988;4:75-87. doi: 10.1007/BF03192102

[31] Kawamura O. Production of monoclonal antibodies against deoxynivalenol and development of a high sensitive ELISA using enzyme amplification. Technical Bulletin of Faculty of Agriculture, Kagawa University. 2005;57:27-33.

[32] Kawamua O, Sato S, Kajii H, Nagayama S, Ohtani K, Chiba J, Ueno Y. A sensitive Enzyme-Linked Immunosorbent assay of Ochratoxin A based on Monoclonal Antibodies. Toxicon. 1989;27:887-897.

[33] Iijima K, Kawamura O, Wang D-S, Manabe M, Tanaka K, Chen G, Yu S-Z, Ueno Y. Development of highly sensitive enzyme-linked immunosorbent assay for fumonisins, and its application for contaminated corn. Mycotoxins. 1996;42:63-66.

[34] Kawamura O, Emoto A. Production of monoclonal antibodies against zearalenone. Technical Bulletin of Faculty of Agriculture, Kagawa University. 2006;58:7-12.

[35] Nagata, S, Soutome H, Tsutsumi T, Hasegawa A, Sekijima M, Sugamata M, Harada K-I, Suganuma M, Ueno Y. Novel monoclonal antibodies against microcystin and their protective activity for hepatotoxixity. Natural Toxins. 1995;3:78-86.

[36] Tabuchi Y, Takabayashi-Yamashita CR, Miguel TA, Ishikawa NK, Maciel LF, Harada K-I, Hirooka EY, Kawamura O. Generation of Monoclonal Antibodies Producing Hybridomas for Detection of Microcystins in Environmental Water by ic-ELISA. In: International Conference on the Water Crisis in the Asia-Pacific Region; 12 February 2015; Kagawa, Japan.

[37] Kamogae M, Hashimoto EH, Pádua CG, Soares FS, Bracarense APFL, Yabe MJS, Ono EYS, Bittencourt-Oliveira MC, Sant`ana CL, Tsutumi T, Nagata S, Ueno Y, Harada H-I, Hirooka EY. Monitoring microcystin and physicochemical parameters: role of agricultural-aquaculture activity in the water quality. In: Njapau H, Trujillo S, Egmond H P V, Park D L, editors. Mycotoxins and phyotoxins: advances in determination, toxicology and exposure management. Netherlands: Wageningen Academic Publishers; 2006. p.321-331.

[38] Visconti A, Pascale M, Centonze G. Determination of ochratoxin A in wine by means of immunoaffinity column clean-up and high-performance liquid chromatography. Journal of Chromatography A. 1999;864: 89-101. DOI: 10.1016/S0021-9673(99)00996-6.

[39] Flajs D, Domijan A. M, Ivic D, Cvjetkovic B, Peraica M. ELISA and HPLC analysis of ochratoxin A in red wines of Croatia. Food Control. 2009;20:590-592. DOI: 10.1016/j.foodcont.2008.08.021

[40] Ogunjimi AA, Choudary PV. Adsorption of endogenous polyphenols relieves the inhibition by fruit juices and fresh produce of immune-PCR detection of Escherichia coli 0157:H7. FEMS Immunology and Medical Microbiology. 1999;23:213-220. DOI: 10.1111/j.1574-695X.1999.tb01241.x.

[41] Agência Nacional de Vigilância Sanitária (ANVISA). 2011. Resolution of Board of Directors- RDC 07/2011, Technical Regulation on Maximum Tolerated for Mycotoxins in Food. Brazil.

[42] Brasil, Diário Oficial da União. 2013. Resolution- RDC n 59, de 26 de dezembro de Dec. 26, 2013, nº 252.

[43] World Health Organization. (WHO). 1998. Guidelines for drinking water quality. 2ª ed Addendum to Volume 2. Geneva.

[44] European Community (EC). 2003. European Commission - Commission Directive 2003/100/EC of 31 October of 2003 amending Annex I to Directive 2002/32/EC of the European Parliament and of the Council on undesirable substances in animal feed. Official Journal of the European Union, L 285, 33-37.

[45] Souza MLM, Sulyok M, Silva OF, Costa SS, Brabet C, Machinski Junior M, Sekiyama BL, Vargas EA, Krska R, Schuhmacher R. Coocurrence of mycotoxins in maize and poultry feeds from Brazil by Liquid Chromatography/Tandem Mass Spectometry. The Sccientific World Journal. 2013;1-9. DOI: http://dx.doi.org/10.1155/2013/427369

[46] Associação Nacional das Indústrias de Biscoitos (ANIB). Dados estatísticos. Available from: http://www.anib.com.br/dados_estatisticos.asp. [Accessed 2013/11/04].

[47] Associação Brasileira da Indústria do Trigo (ABITRIGO). Evolução mensal e anual do preço do trigo, consumo, produção e estoque mundial de trigo. Available from: http://www.abitrigo.com.br. [Accessed 2013/10/31].

[48] Associação Brasileira da Indústria do Trigo (ABITRIGO). Available from: http://www.abitrigo.com.br. [Accessed 2015/02/26].

[49] Associação Brasileira das Indústrias de Biscoitos, Massas Alimentícias e Pães & Bolos Industrializados (ABIMAPI). Available from: http://abima.com.br/estatistica-biscoito.php. [Accessed 2015/02/26].

[50] Almeida-Ferreira G, Barbosa-Tessmann IPB, Sega R, Machinski Junior M. Occurence of zearalenone in wheat- and corn- based products commercialized in the State of Paraná, Brazil. Brazilian Journal of Microbiology. 2013;44:371-375. DOI: 10.1590/S1517-83822013005000037

[51] Magan N, Hope R, Colleate A, Baxter ES. Relationship between growth and mycotoxin production by Fusarium species, biocides and environment. European Journal of Plant Pathology. 2002;108:685-690. DOI: 10.1023/A:1020618728175.

[52] Müllenborn C, Steiner U, Ludwig M, Oerke EC. Effect of fungicides on the complex of Fusarium species and saprophytic fungi colonizing wheat kernels. European Journal of Plant Pathology. 2008;120:157-166. DOI: 10.1007/s10658-007-9204-y.

[53] Moss MO, Frank JM. Influence of the fungicide tridemorph on T-2 toxin production by Fusarium sporotrichioides. Transactions of the British Mycological Society. 1985;54:585-590. DOI: 10.1016/S0007-1536(85)80111-X.

[54] Doohan FM, Weston G, Rezanoor HN, Parry DW, Nicholson P. Development and use of a reverse transcription – PCR assay to study expression of tri5 by Fusarium species in vitro and in plant. Applied and Environmental Microbiology. 1999;65:3850-3854.

[55] Falcao VCA, Ono MA, Miguel TA, Vizoni E, Hirooka EY, Ono EYS. Fusarium verticillioides: Evaluation of fumonisin production and effect of fungicides on in vitro inhibition of mycelial growth. Mycopathologia. 2011;170:77-84. DOI: 10.1007/s11046-010-9339-9.

[56] Miguel TA, Bordini JG, Saito GH, Andrade CGTJ, Ono MA, Hirooka EY, Vizoni E, Ono EYS. Effect of fungicide on Fusarium verticillioides mycelial morphology and fumonisin B1 production. Brazilian Journal of Microbiology. 2015;46:xxx-xxx. DOI: http://dx.doi.org/10.1590/S1517-838246120120383

[57] Ochiai N, Fujimura M, Oshima M, Motoyama T, Ichiishi A, Yamada-Okabe H, Yamaguchi I. Effects of iprodione and fludioxonil on glycerol synthesis and hyphal development in Candida albicans. Bioscience, Biotechnology, and Biochemistry. 2002;66:2209-2215. DOI: 10.1271/bbb.66.2209

[58] Figueira ELZ, Blanco-Labra A, Gerage AC, Ono EYS, Mendiola-Olaya E, Ueno Y, Hirooka EY. New amylase inhibitor present in corn seeds active in vitro against amylase from *Fusarium verticillioides*. Plant Disease. 2003;87:233-240. DOI: 10.1094/PDIS.2003.87.3.233.

[59] Moreno EC, Garcia GT, Ono MA, Vizoni E, Kawamura O, Hirooka EY, Ono EYS. Co-occurrence of mycotoxins in corn samples from the Northern region of Paraná State, Brazil. Food Chemistry. 2009;116:220-226. DOI: 10.1016/j.foodchem.2009.02.037.

[60] Ono EYS, Moreno EC, Ono MA, Rossi CN, Saito GH, Vizoni E, Sugiura Y, Hirooka EY. Effect of cropping systems and crop successions on fumonisin levels in corn from Northern Paraná State, Brazil. European Journal of Plant Pathology. 2011;131:653-660. DOI: 10.1007/s10658-011-9839-6.

[61] Marocco A, Gavazzi C, Pietri A, Tabaglio V. On fumonisin incidence in monoculture maize under no-till, conventional tillage and two nitrogen fertilization levels. Journal of the Science of Food and Agriculture. 2008;88:1217-1221. DOI: 10.1002/jsfa.3205.

[62] Blandino M, Reyneri A, Vanara F. Influence of nitrogen fertilization on mycotoxin contamination of maize kernels. Crop Protection. 2008;27:222-230. DOI:10.1016/j.cropro.2007.05.008.

[63] Risso WE. Influence of agricultural practices on the chemical composition and sanitary quality: fumonisin in corn [thesis]. Londrina: State University of Londrina; 2013.

[64] Puri K, Zhang S. The 3ADON population of Fusarium graminearum found in North Dakota is more aggressive and produces a higher level of DON than the prevalent 15ADON population in spring wheat. Phytopathology. 2010;100:1007-1014. DOI: 10.1094/PHYTO-12-09-0332.

[65] Ward TJ, Clear RM, Rooney AP, O'Donnell K, Gaba D, Patrick S, Starkey DE, Gilbert J, Geiser DM, Nowicki TW. An adaptive evolutionary shift in*Fusarium* head blight pathogen populations is driving the rapid spread of more toxigenic *Fusarium graminearum* in North America. Fungal Genetics and Biology. 2008;45:473–484.

[66] Wu Q, Dohnal V, Huang L, Kuča K, Yuan Z. Metabolic pathways of trichothecenes. Drug Metabolism Reviews. 2010; 42:250–267.

[67] Valério F, Favilla M, De Bellis P, Sisto A, De Candia S, Lavermicocca P. Antifungal activity of strains of lactic acid bacteria isolated from a semolina ecosystem against Penicillium roqueforti, Aspergillus niger and Endomyces fibuliger contaminating bakery products. Systematic annd Applied Microbiology. 2009;32:438–448. DOI: 10.1016/j.syapm.2009.01.004.

[68] Franco TS, Garcia S, Hirooka EY, Ono YS, dos Santos JS. Lactic acid bacteria in the inhibition of Fusarium graminearum and deoxynivalenol detoxification. Journal of Applied Microbiology. 2011;111(3):739-48. DOI: 10.1111/j.1365-2672.2011.05074.x.

[69] Zhang XB, Ohta Y. Binding of mutagens by fractions of lactic acid bacteria on mutagens the cell wall skeleton. Journal of Dairy Science. 1991;74:1477–1481.

[70] Niderkorn V, Boudra H, Morgavi DP. Binding of Fusarium mycotoxins by fermentative bacteria in vitro. Journal of Applied Microbiology. 2006;101:849–856. DOI: 10.1111j.1365.2672.2006.02958.x.

[71] Mozzi F, Raya RR, Vignolo GM. Editors. Biotechnology of lactic acid bacteria :novel applications. Blackwell Publishing. 2010.

[72] Orman BA, Schumann Jr RA. Comparison of near-infrared spectroscopy calibration methods for the prediction of protein, oil and starch in maize grain. Journal of Agricultural and Food Chemistry. 1991;39:883–886. DOI: 10.1021/jf00005a015.

[73] Armstrong PR. Rapid single-kernel NIR measurement of grain and oil-seed attribute. Applied Engineering in agricultura. 2006;22:767-772. DOI: 10.13031/2013.21991.

[74] Fox G, Manley M. Applications of single kernel conventional and hyperspectral imaging near infrared spectroscopy in cereals. Journal of the Science of Food and Agriculture. 2013;94:174-179. DOI: 10.1002/jsfa.6367.

[75] Wang W, Heitschmidt GW, Ni X, Windham WR, Hawkins S, Chu X. Identification of aflatoxin B_1 on maize kernel surfaces using hyperspectral imaging. Food Control. 2014;42:78-86. DOI: 10.1016/j.foodcont.2014.01.038.

[76] Berardo N, Pisacane V, Battilani P, Scandolara A, Pietri A, Marocco A. Rapid Detection of Kernel Rots and Mycotoxins in Maize by Near-Infrared Reflectance Spectroscopy. Journal of Agricultural and Food Chemistry. 2005;53:8128-8134. DOI: 10.1021/jf0512297.

[77] Barbin DF, Kaminishikawahara CM, Soares AL, Mizubuti IY, Grespan M, Shimokomaki M, Hirooka EY. Prediction of chicken quality attributes by near infrared spectroscopy. Food Chemistry. 2015;168:554–560. DOI: 10.1016/j.foodchem.2014.07.101.

[78] Agrostat Brasil. 2015. Sistema de Estatísticas de Comércio Exterior do Agronegócio. Available from: http://sistemasweb.agricultura.gov.br/pages/AGROSTAT.html [Accessed: 2015/02/25].

[79] Food and Agriculture Organization of the United Nations (FAO). 2015. Statistics division of FAO – FAOSTAT. Available from: http://faostat.fao.org/site/339/default.aspx. [Accessed: 2015/02/26].

[80] Brazil. Ministério da Agricultura, Pecuária e Abastecimento. 2013. Projecões do Agronegócio: Brasil 2012/2013 a 2022/2023 / Ministério da Agricultura, Pecuária e Abastecimento. Assessoria de Gestão Estratégica.– Brasília : Mapa/ACS, 2013. 96 p.

[81] Ministry of Finance, Zaimusho - Japan. ALIC - Agriculture & Livestock Industries Corporation. Available from: http://www.alic.go.jp/joho-c/joho05_000073.html. [Accessed: 2015/01/26].

Chapter 6

Low-Fat Foods

Suzana Caetano da Silva Lannes and
Alline Aurea do Amaral

Additional information is available at the end of the chapter

http://dx.doi.org/10.5772/60817

Abstract

Fats are responsible for performing varied and important functions in the body, such as providing calories, essential fatty acids, and fat-soluble vitamins. They are considered very important among the ingredients and the sensory aspects of the functional properties of the food. They influence the melting point, consistency, and formation of crystals in the spreadability of many foods and are also responsible for flavor, aroma, creamy appearance, aeration, stability, and feeling of fullness after meals. However, the consumption of high amounts of fats and oils has often been associated with obesity and multiple chronic diseases. To reduce fat and caloric value of foods, we can reduce or eliminate fat from the formulation by increasing the amount of proteins, carbohydrates, fibers, and water. However, it is not so easy to treat with fat substitution in a food formulation. The crystallization behavior of lipids has important implications, especially in industrial processing of products whose physical characteristics (consistency and melting point) are affected by the crystal structure of fat, such as chocolate, margarine, and shortenings. Much of the knowledge about the crystal structure of the fat comes from studies performed on diffraction of x-rays. The crystal structure depends on the specific type of triacylglycerol (TAG) present, the composition and distribution of fatty acids, the purity of TAG, and the crystallization conditions (temperature, cooling rate, shear, and solvent). The ideal fat replacers should be a composite of recognized safety and health, which has all the functional and organoleptic properties with the benefit of significant calorie reduction. Fat food processors have been careful about developing and producing low-fat foods due to the problems it could generate in the production, such as an increased risk of unstable products and unconfident production parameters. Some fat foods can be considered as margarines, creams, chocolate products, ice creams, cakes, and some baked goods.

The substitution of *trans* and saturated fatty acids should be considered also when creating a low-fat food. Saturated and *trans* fats refer to a group of fatty acids, each with its own properties and characteristics. Despite saturated fats' potential health benefits, saturated fat has long been associated with increased risk of heart disease, stroke, and even cancer, as well *trans* fats. When designing a low-fat food as spread, for example, it is important to observe the fat composition, the quantity of liquid oil, because the oil phase needs to cover a higher amount of water droplets and the solid fats cause disappearance of smoothness. A combination of the right process parameters and fat composition can give a satisfactory fat food product. In this chapter, the possibilities of low-fat food creation are discussed.

Keywords: Fats and oils, food structure, food processing, food health

1. Introduction

The final characteristics of processed fat products depend on the physical and chemical properties of oils and fats present in their formulation. To obtain the required specifications for each product, different fatty bases have to be formulated. The knowledge of physical, rheological, chemical, and sensory characteristics, functionality, and fat interactions with other ingredients is essential for formulating these bases.

Vegetable oil and vegetable fats are products consisting primarily of glycerides of fatty acids found in different types of plants. They may also contain small amounts of other lipids, such as phospholipids, and unsaponificable constituents and free fatty acids naturally present in oil or fats. Vegetable oils are liquid at 25°C, and vegetable fats are solid or pasty at 25°C.

There are different types of vegetable oils used by the industry to formulate fat bases: soybean, cotton, peanut, sunflower, canola, sesame, corn, olive, palm, palm kernel, coconut, cocoa, linseed, and castor oil, as well as oils and fats obtained from fish, beef, pork, and poultry.

According to Brazilian laws and regulations [1], vegetable fats are derived from various sources and defined as products made primarily of glycerol of fatty acids found in plant species. Chemically, all oils and fats are considered triacylglicerols or esters of glycerol and fatty acids, which are responsible for the different properties observed in these molecules due to their size, saturation, and/or position. When comparing chains of the same length, saturated structures are less reactive than those with unsaturation [2].

The importance of fats for humans, animals, and plants is their energy content (9 kcal/g). It is a source of fat-soluble vitamins (A, D, E, and K) and of essential fatty acids (omegas 3 and 6); they act as heat transfer medium and contribute to texture, flavor, and color of foods.

Fat substitutes (fat replacers) can replace fat in food products; however, they often change texture and/or flavor of foods or beverages. Partial replacement of fat is generally a better approach in terms of consumer acceptance.

http://dx.doi.org/10.5772/60817

It was in the 1980s that consumers became aware of impact of diet on health; it was then proposed a reduction to 30% (from 40% to 49%) of energy from fat in diet, which started to affect consumer attitudes. The challenge was to produce low-fat products with physical and sensory characteristics as close as possible to full-fat quality.

Protein-based fat substitutes came along with the introduction of a microparticulated protein product called *Simplesse*. Subsequent development efforts revealed consequences of removing fat from a product. Alternative ingredients or processes had to be developed as all the attributes of fat became recognized.

2. Functionality and crystallization of fats

Lipids are a family of organic compounds soluble in organic solvents but not in water. The lipid class can be divided into the following categories: triacylglicerols (95% of lipids in foods), phospholipids such as lecithin, and sterols such as phytosterol and cholesterol.

Sensory functions of fats in foods can be as follows: appearance (gloss, translucency, color, surface uniformity, and crystallinity), texture (viscosity, elasticity, and hardness), flavor and aroma (flavor release profile and development), and mouthfeel (meltability, creaminess, thickness, degree of mouthcoating, and mouth warming or cooling).

Unsaturated fatty acids may have *cis* and *trans* configurations with different physicochemical properties. Due to their structural characteristics, fatty acids in the *trans* configuration have higher melting point than its corresponding *cis* isomer but lower than the melting point of the saturated fatty acid with an even number of carbon atoms. Thus, the *trans* isomer can be considered as intermediate between an original unsaturated fatty acid and a completely saturated fatty acid [2].

The crystallization behavior of lipids has important implications, especially in industrial processing of products whose physical characteristics (consistency and melting point) are affected by the crystal structure of the fat, such as chocolate, margarine, and shortenings. Most of the knowledge about the crystal structure of the fat comes from studies done with X-ray diffraction. The structure of the crystal depends on the specific type of triacylglycerol (TAG) present, the composition and distribution of fatty acids, the purity of TAG, and the crystallization conditions (temperature, cooling rate, shear, and solvent) [3, 4].

The crystallization process consists of two events: nucleation and crystal growth. Nucleation involves the formation of aggregates of molecules that exceed a critical size and are therefore stable [5, 6]. The three polymorphic forms of fat crystals, in order of increasing thermodynamic stability, are alpha (α), beta press (β'), and beta (β); the choice of fat should be based on the polymorph β' to promote excellent creaminess properties [7].

Lipid crystallization behavior has important implications, especially in industrial processing of products whose physical characteristics depend largely of fat crystals, such as chocolates, margarine, and shortenings.

The crystal growth rate and the polymorphic transformations are important to determine the process and conditions of storage of oils and fats.

Plastic fats consist of a crystal network in the continuous oil matrix. The rheological behavior of these fats is the result of interactions between the crystals that are immersed in this liquid matrix.

The liquid portion, in conjunction with the solid fraction, is responsible for the viscoelastic behavior of a plastic fat. Thus, the amount of crystallized fat and the type of fat matrix crystals are of paramount importance in the rheological behavior of the fat.

The main objectives of fat crystallization are to increase the thermal stability and oxidative stability of the product, to increase plasticity, to promote creaminess and smoothness, to standardize of the physical and chemical characteristics of the product, and to guarantee performance.

The structure of fat crystals depends on the conditions of its formation, such as stirring, cooling rate, and fat quenching, leading to a specific consistency (plasticity).

Some important properties of acyl chains are providing flexibility that consequently leads to a molecular conformation that results in lateral packing and their aggregation state due to the solid and liquid crystalline states created. The physical properties of long-chain compounds in the solid state, such as melting point, heat competence, and plasticity, depend on the crystal structure and are very important to the industry. TAGs exhibit various crystalline forms. The solid states are related to the quality of industrial products. Heat treatment is used in the production processing of oil and fat products. Information about crystal structures of fats and complex lipids can be provided by X-ray diffraction analysis [8].

3. Fats versus health

More and more people are trying to lose weight. In general, people continue to understand that restrictive dieting (deprivation, short-term solutions) spells failure. Instead, it takes permanent lifestyle changes to lose weight and prevent weight gain. Fifty-four percent of U.S. adults are currently trying to reduce their weight. (Note: previous surveys asked about "dieting," but in 2010, the survey was modified to refer to weight reduction or "trying to lose weight" [9].)

However, the consumption of high amounts of oils and fats has often been associated with obesity and various types of chronic diseases. The reduction of fat and caloric value of foods can reduce or eliminate fat from the formulation increasing the protein, carbohydrate, fiber, and water content.

High-fat diets are associated with obesity and cardiovascular disease. There is an epidemic of overweight and obesity in some locations. Individuals with BMI $\geq$ 25 (BMI = weight (kg)/height ^{2} (m^{2}) are considered overweight, and with BMI $\geq$ 30 they are considered obese. The recom-

http://dx.doi.org/10.5772/60817

mended dietary allowance (RDA) intake of fat is 30%, but the actual consumption is higher: calories, 9 kcal/gram; carbohydrates and protein, 4 kcal/g of.

4. Fat substitutes

An ideal fat substitute should be a compound of recognized as safe and which has all the functional and organoleptic properties and significantly less fat and calories. Due to the multiple roles played by fats in determining product properties (thermal stability, emulsification and aeration, lubrication, taste, and mouthfeel), it is usually necessary to use a combination of different fat substitutes in order to have products with these expected properties [10].

Fat replacers can be classified into two groups according to their mechanism of action: fat mimetics or derivatives. They can also be classified into three groups based on their chemical nature: carbohydrates, proteins, or fats.

According to the mechanism of action, the fat mimetics are a combination of water and lipids or nonlipids (modified proteins or carbohydrates) and emulsifying properties; they are capable of forming gels and thus normally serve as body agents, for example, cellulose, gum, inulin, maltodextrin, starches and modified starches, proteins, and microparticulate concentrated protein. Derivatives are noncaloric or low-calorie fat compounds with properties similar to those lipids, which ester bonds are modified.

According to the Calorie Control Council [9], fat replacers can be classified and exemplified as follows:

Type	Commercial name	Source	Calories	Application
Microparticulated protein	Simplesse®	Whey protein or milk and egg protein	1–2 kcal/g	Dairy products, salad dressing, margarine, mayonnaise-type products, bakery
Modified whey protein concentrate	Dairy-Lo®	Protein	Reduced calorie	Milk/dairy products, baked goods, salad dressing, mayonnaise-type products
Similar to microparticulated protein but made by a different process	K-Blazer®, ULTRA-BAKE™, ULTRA-FREEZE™, Lita®	Based on egg white and milk proteins or derived from a corn protein	Reduced calorie fat substitute and fat replacer	Frozen desserts, bakery

Table 1. Protein-based fat replacers

Type	Commercial name	Source	Calories	Application
Cellulose	Avicel® cellulose gel, Methocel™, Solka-Floc®	Purified form of cellulose ground to microparticles	Noncaloric	Dairy-type products, sauces, frozen desserts, salad dressings
Dextrins	Amylum, N-Oil®	Dextrins, tapioca	4 kcal/g	Salad dressings, puddings, spreads, dairy-type products, frozen desserts
Fiber	Opta™, Oat Fiber, Snowite, Ultracel™, Z-Trim	Fiber	Low caloric	Bakery, meats, spreads, extruded products
Gums	KELCOGEL®, KELTROL®, Slendid™	Guar gum, gum arabic, locust bean gum, xanthan gum, carrageenan, pectin	Noncaloric	Salad dressings, desserts, and processed meats
Inulin	Raftiline®, Fruitafit®, Fibruline®	Fiber and bulking agent extracted from chicory root	1–1.2 kcal/g	Yogurt, cheese, frozen desserts, bakery, icings, fillings, whipped cream, dairy products, processed meats
Maltodextrins	CrystaLean®, Lorelite, Lycadex®, MALTRIN®, Paselli®D-LITE, Paselli®EXCEL, Paselli®SA2, STAR-DRI®	Carbohydrate sources such as corn, potato, wheat, and tapioca	4 kcal/g	Bakery, dairy products, salad dressings, spreads, fillings, processed meat, frozen desserts, extruded products, beverages
Nu-Trim		From oat and barley	Reduced calorie	Bakery, milk, cheese, ice cream
Oatrim	Beta-Trim™, TrimChoice	Hydrolyzed oat flour	1–4 kcal/g	Bakery, fillings, frozen desserts, dairy beverages, cheese, salad dressings, processed meats and confectionery
Polydextrose	Litesse®, Sta-Lite™	From dextrose containing minor amounts of sorbitol and citric acid	1 kcal/g	Baked goods, chewing gums, confectionery, salad dressings, frozen dairy desserts, gelatins, puddings

http://dx.doi.org/10.5772/60817

Type	Commercial name	Source	Calories	Application
Polyols	Many brands available		1.6–3.0 kcal/g	Reduced fat and fat-free products
Starch and modified food starch	Amalean®I & II, Fairnex™VA15, & VA20, Instant Stellar™, N-Lite, OptaGrade®#, Perfectamyl™AC, AX-1, & AX-2, PURE-GEL®, STA-SLIM™	From potato, corn, oat, rice, wheat, or tapioca starches	1–4 kcal/g	Processed meats, salad dressings, bakery, fillings, sauces, condiments, frozen desserts, dairy products
Z-Trim	Appears as cornstarch on the ingredient statement, others appear as food starch modified	Made from insoluble fiber from oat, soybean, pea, and rice hulls or corn or wheat bran	Calorie free	Bakery, burgers, hot dogs, cheese, ice cream, yogurt

Table 2. Carbohydrate-based fat replacers

Type	Commercial name	Source	Calories	Application
Olestra	Olean®	Made from sucrose and edible fats and oils	Calorie-free	Salty snacks, crackers, frying
Sorbestrin	*May require FDA approval*	Liquid fat substitute composed of fatty acid esters of sorbitol and sorbitol anhydrides	Low-calorie (approximately 1.5 kcal/g)	Fried foods, salad dressing, mayonnaise, bakery
Esterified propoxylated glycerol (EPG)	*May require FDA approval*		Reduced-calorie	All typical consumer and commercial applications, bakery, frying

Table 3. Lipid (Fat/Oil) analogs

Type	Commercial name	Source	Calories	Application
Emulsifiers	Dur-Lo®, EC^{T}-25	Vegetable oil mono- and diglyceride + water, sucrose fatty acid esters, emulsion systems using soybean oil or milk fat	9 kcal/g	Mixes, cookies, icings, vegetable dairy products
Salatrim	Benefat^{T}	Structured lipid (Short And Long Acyl Triglicéride Molecule)	5 kcal/g	Bakery, confectionery

Table 4. Fat-based fat replacers

Using small amounts of inulin, it is possible to improve the flavor and texture of products with low fat content. Combinations with polyols or fructose give excellent results in chocolates.

Plum products can be used to replace fat in products with reduced fat content due to the presence of sorbitol and fiber, which can bind the water. These products have been used as fat substitute in bakery products and confectionery [11].

Organogels are mixtures with characteristics of gel, consisting of structuring agents and immobilized phase forming a thermoreversible self-sustaining network, in which the immobilized phase is an organic compound, which differs from other gels formed primarily of water-soluble compounds. The main structuring agents that have been studied for the formation of organogels are diacylglycerols, monoacylglycerols, fatty acids and alcohols and mixtures thereof, waxes and wax esters, phytosterol and oryzanol, lecithin, and sorbitan tristearate, among others. The use of organogels in food products has been the subject of several studies as an alternative to obtain products with appropriate technological properties without the use of *trans* fats and reduced saturated fatty acid content [12].

4.1. Hydrocolloids

Hydrocolloids are substances that form colloidal systems when dispersed in water. They are polymers of carbohydrates (also proteins) and are derived from a wide range of natural sources or are produced synthetically. Hydrocolloids are widely exploited in industry, not only in the food sector, for their ability to control important functional properties, including thickening and gelling, stabilization, dispersion, and emulsification.

In processed foods, hydrocolloids are responsible for texture control and stabilization of these products, and therefore they can prevent or retard a number of physical phenomena such as sedimentation of solid particles suspended in the medium (water) or the crystallization of the sugar or disaggregation and aggregation of dispersed particles [13]. In products with total or partial fat reduction, gums minimize texture changes and prevent phase separation in emulsions [14].

Gums can replace fats in certain formulations since they have the ability to swell or solubilize in aqueous systems providing viscosity characteristics similar to those of fats [15]. In addition, the substitution of fat for ingredients with low calorie content has revolutionized the food industry due to the demand of consumers concerned about health. This factor has allowed many technological advances, and improvements in the functionality and palatability of these products have been made. The addition of starch in foods can also provide nutritional benefits associated with dietary fiber intake [16, 17].

The use of natural gums in the food industry is considered safe, according to the Food and Drug Administration (FDA) [18]; the domestic production of some natural polysaccharides as an alternative to existing products or for the manufacture of new products is a possibility [15, 19].

It was possible the development of chocolate spread with the addition of gelatin in partial replacement of fat, as evaluation of control (with vegetable fat) and with addition of solution of gelatin (1.0%) from chicken feet to replace 50% of the fat [20].

Three conventional fat-based formulations for filling were prepared with low *trans* fat, at concentrations of 10%, 7.5%, and 5%, and six others with the addition of 0.3% (w/v) of guar gum and chichá gum individually. Rheological study showed the formulations made with guar and chichá gums pointed similar behavior and pronounced thixotropy by a decrease in apparent viscosity with time, followed by recovery of the structure of the system. Thus, it is concluded that the rheological properties of the fillings are significantly altered by the addition of the studied gums, substituting part of the fat, even at low concentrations [21].

Light mayonnaise was obtained, produced from 25.5% starch (+24.82% oil) and 25.5% hydrogel chitosan (+34.50% oil). Chitosan leads mayonnaise to a lighter texture. Mayonnaise, known to be as calorie sauce, had reduced oil content in addition the biological, nutritional, and chemical properties of the biopolymer chitosan [22].

Reduced fat and egg mayonnaise samples were produced with different fat replacers (xanthan, guar, and pregelatinized cornstarch) and egg/soy milk mixture as egg alternative, concluding that due to the capability of xanthan and pregel cornstarch in changing physicochemical parameters, they can be used in mayonnaise and other food formulations as fat replacer [23].

4.2. Structured lipids

Structured lipids are TAGs restructured or modified to change the fatty acid composition and/or their positional distribution in glycerol molecules by a chemical or enzymatic process. Salatrim is a typical example in this category.

They can be synthesized in order to improve or modify the physical (polymorphism, melting point, solid fat content, viscosity, and consistency) and/or chemical (oxidative stability) characteristics of triglycerides, as well as to modify one or more of their nutritional properties (presence or absence of saturated fatty acids or unsaturated fatty acids easily absorbed and digested).

Some benefits of specially formulated lipids are as follows: they can improve immune function, decrease risk of cancer, prevent stroke, and decrease blood cholesterol levels, and they have low calories.

4.3. Cocoa butter

Since the 1930s, there has been an increased interest in using fats other than cocoa butter in confectionery production due to the uncertainty of supply and costs of cocoa butter, which are dependent on the fluctuating cocoa bean market.

When a different fat composition is added to cocoa butter, the resulting crystalline form of the fat is typically altered, producing changes in the melting profile of the fat, called incompatibility.

To replace cocoa butter completely or partially, industry and researchers have attempted to develop substitutes by producing fats with features that meet consumer demands. These are classified as follows [24, 25]:

a. Substitutes (cocoa butter substitutes [CBS])—similar to cocoa butter in its physical properties, but not fully compatible for mixtures. Can be divided into lauric and nonlauric fats

b. Equivalent (cocoa butter equivalents [CBE])—similar in physical and chemical properties and can be used in mixtures in any proportion since they contain almost the same type of fatty acids and glycerides of cocoa butter

The stearic acid has a unique position within the saturated fatty acids of long chain. It is shown that, unlike other saturated fatty acids, it has a neutral cholesterolemic effect.

Cocoa butter contains lots of saturated fatty acids in the TAG positions, which are less easily absorbed and thus do not reduce the risk of atherosclerosis.

Some details, when cocoa butter is substituted, are its technological and nutritional characteristics. For example, the use of special oils in creamy fillings must be compatible with the fat coating and migration of the filling towards the coating may occur, or vice versa. This process can affect the integrity and appearance of the product.

4.4. Nutritional and functional shortenings

Industries produce fats for many applications, and some nutritional benefits are expected. Saturate All Purpose and Emulsified Shortening are products free of partially hydrogenated oil and have reduced content in saturated fat compared to traditional shortening, polyunsaturated, and monounsaturated fat products; they have zero grams *trans* fat per serving.

Saturated fats can cause cardiovascular diseases, hypertension, obesity, diabetes mellitus, inflammatory and autoimmune disorders, cancer, increased blood cholesterol levels, and gallbladder diseases.

Monounsaturated fats are considered good fats because they help protect the heart against cardiovascular diseases. A benefit of polyunsaturated fatty acids is that they reduce the triglyceride levels. It is recommended to substitute saturated fat in the diet for polyunsaturated fats. Polyunsaturated fats attach to and clear out unhealthy fats, such as saturated fat, cholesterol, and triglycerides. Higher intake of most dietary saturated fatty acids is associated with higher levels of total blood cholesterol and low-density lipoprotein (LDL) cholesterol. High total and LDL cholesterol levels and high intake of *trans* fatty acid are risk factors for cardiovascular disease.

It is suggested the selection of features and beneficial properties of lipids and exclusion (or decrease) of fats (or its components) that cause health problems.

4.5. *Trans* and low *trans* fats

The characteristics of a low *trans* fat are as follows: acceptable level of *trans* fatty acids, acceptable level of saturated fatty acids, palm oil use, application/performance, shelf life required. From a technological point of view, the best type of fat has fast crystallization and beta prime crystal form. Alternative low *trans* fats are palm, corn, and cotton oil (little or nonhydrogenated oils). Other alternatives are palm kernel, coconut, high oleic canola, medium-oleic sunflower, high oleic soybean, and low linolenic soybean oil and also animal fats.

There are a number of challenges faced by the food industry during the process of developing alternatives to low *trans* (LT)/low saturated (LS) fats such as the replacement of oils and fats seeking an alternative to LT/LS while keeping the same functionality in the final product. Another challenge is the availability of these oils and fats. The technologies to obtain LT and LS fats may be used alone or combined: mixtures of different sources of oils and fats, interesterification, hydrogenation, fractionation, and genetics. For example, high oleic canola oil + fully hydrogenated soybean oil (70°C) = interesterified fat (melting temperature, 35°C–40°C).

The change to zero *trans* fat can modify the product capability, performance, oxidative stability, flavor, and color.

Trans fats that contain *trans* fatty acids may be produced by either natural or industrial hydrogenation. Biohydrogenation occurs when fatty acids ingested by ruminants are partially hydrogenated by enzymatic systems of the intestinal tract of the animals. Industrial hydrogenation is the addition of nitrogen polyunsaturated vegetable oils using a catalyst under appropriate temperature and pressure. This process promotes the formation of fatty acids with high melting point due to the linear guidance molecules in *trans* fats and the increased saturation index and increased stability to lipid oxidation process [26].

The hydrogenated vegetable fat is responsible for most of the consumption of *trans* fatty acids in the diet. In Brazil, the commercial hydrogenation of vegetable oils date from the 1950s, and it aims at producing technical fats (shortenings), margarines, and fats for frying. With the development of selective hydrogenation techniques, processed vegetable oils quickly replaced animal fats in the diet of Brazilians. These fats have been widely used in the production of various foods such as margarine, chocolate icing, cookies, baked goods, ice cream, pasta, and potato chips, among others [27].

In the past, the formation of *trans* isomers was considered as a technological advantage since, due to its higher melting point compared to that of the corresponding *cis*, it facilitated the creation of solid desirable levels of hydrogenated fat [28, 29].

Scientific evidences related to the negative impact of *trans* fatty acids on health (heart disease risk, high LDL cholesterol levels, low HDL levels) have led to minimized consumption of partially hydrogenated fat, fact corroborated by progressive changes in law and regulations in several countries. The challenge of the food industry in terms of *trans* fat replacement in various products is to develop formulations with comparable economic viability and functionality, and which do not result, however, in a substantial increase in the saturated fatty acid content in foods [30].

The low *trans* fats that are free of *trans* fatty acids, for example, obtained by interesterification, have been used to replace hydrogenated vegetable fat. The interesterification of oils and fats may be chemical or enzymatic. In this process, the fatty acids are redistributed in ester bonds, creating new groups without the formation of *trans*-isomers. The chemical interesterification process has long been used for its ease and low cost, and it is usually carried out with mixtures of highly saturated fats and liquid oils [2]. In general, blends of coconut and palm oils, after modifications, have great potential in the development of healthier fat products. Enzymatic interesterification after a full hydrogenation is useful in the production of low *trans* fats.

The application of most oils and fats in their natural form in food products is very limited due to their physicochemical properties. However, by increasing their functionality and nutritional value, their behavior can be changed. Adaptation of the melting profile increases stability and shelf life.

The most difficult challenge is to replace *trans* fat in shortening. Solid fats are desirable in foods. Saturated fats have been replaced with hydrogenated (*trans*) fats due to health issues associated to fat, but now, saturated fat is also desirable again.

Trans fatty acids can be replaced, although it is very difficult to remove these fats from foods. However, success has been obtained with significant research efforts.

5. Fat and low-fat products

A food product contains two or more of the following ingredients: carbohydrates, proteins, fats, water, and additives (colorants, flavors). Each ingredient has a function in the product. Fat is the most abundant ingredient in fatty or high-fat products; it has some functions in foodstuff such as: aeration emulsifying properties, improving texture, flavor releasing, facilitating the mixing process, and staining.

The correct product structure is essential for making products with desirable properties. The product structure is then dependent on the type and the amount of fat present in its formulation. The chemical properties of oils and fats (carbon chain length, degree of unsaturation, distribution of fatty acids, *cis–trans* configuration, and crystalline state) are responsible for the physical properties of the food, i.e., their structure, properties, and rheological and thermal behavior during processing and post-processing, and stability during storage.

In many products, the outward appearance is defined by the macrostructure and determined by processing. The important factor in innovations of novel food products is the development of new process concepts dedicated to food structuring purposes. Particles of emulsions, foam bubbles, fat globules serve a microscale, which is related to the structure of the product (1–100 μm).

5.1. Low-fat products

According to the American Cancer Society [31], foods like margarine and mayonnaise with a high fat content must have half or less than half the fat of the regular version of the food to be called *light*. These foods do not usually meet the 30% cutoff for number of calories from fat to be considered low fat. Low-fat foods can be, for example, dairy and dairy-like products (low-fat (1%), fat-free (skim) yogurt, cottage cheese, milk, sorbet, sherbet, gelatin ices, and low-fat

http://dx.doi.org/10.5772/60817

or fat-free frozen yogurt); fish, meat, and poultry (egg whites or egg substitutes, crab, white fish, shrimp, veggie burgers); grains, cereals (except granola types), and pastas (rice or noodles (watch out for fat in sauces you may add); bagels, pita bread, or English muffins; low-fat crackers, soft tortillas, toast, muffins, crackers, or plain breads); fruits and vegetables (fruit juices, vegetable juices); and sweets (gelatin, low-fat microwave popcorn, hard and jelly candies).

Diet chocolate is known to be a product rich in fat. Its caloric value can decrease with a decrease in the concentration of fat. However, when the fat content is less than 27% of its weight, the chocolate loses its softness and its melting in the mouth. The main ingredients of diet chocolate, in addition to the traditionally used ingredients (such as cocoa mass, cocoa butter, milk powder, and others), are monitol, maltitol, xylitol, sorbitol, and fructose. The use of anhydrous crystalline forms of polyols such as isomalt, maltitol, and lactitol facilitates the process [32].

5.2. Functionality of fat substitutes

Proteins are good substitutes for fat (structuring of the aqueous phase) function, for example, as a gelling agent or thickener. The properties of gels depend on the type and concentration of protein, pH, type, and concentration of salt and heat treatment. These also have the water and fat holding capacity, as well as air retention.

Polysaccharides also have water retention capacity (network formation), serving as a thickener (high molecular weight and high capacity to bind water), stabilizing and gelling.

Viscoelastic properties of dilute solutions of polymer materials such as carbohydrates and proteins can be used to characterize their three dimensional configuration in the solution, that can affect its functionality in many food products. It is possible to improve the food flow behavior by understanding how the polymer molecular structure affects its flow properties, for example, improving the consistency and stability of emulsions using polymers with increased surface area and increased viscosity and elasticity.

In a mayonnaise-like emulsion, which is a viscoelastic material, these features are associated with complex interactions between oil droplets and biopolymers present (egg protein and polysaccharide). The yolk lipoproteins influence the viscoelasticity.

In designing frozen food structure, there is a high dependence between product quality and structure of crystals, which is defined by size, number, shape, and arrangement of crystals. The structure of the crystals formed (ice, fat, and sugar crystals) is influenced by process parameters and ingredients. High variation of temperature during storage can change and increase the size of crystals. Proteins and hydrocolloids (as antifreeze agents) can be added to formulations to create rectangular and elongate ice crystals, which can lead to a high quality product.

Fats and sugars influence mechanical properties, melting resistance, and palatability of ice creams. There are technological effects on different sugars (fructose syrup in blends with sucrose instead glucose syrup) and fats (palm fat as an alternative to hydrogenated vegetable) on the structure of ice cream formulations. Hygroscopicity of fructose syrup increased the solids content in the formulations. Melting rate and overrun were higher in products added with fructose syrup. Palm fat caused changes in melting ranges of formulations, and higher melting rate was observed in the combination of palm fat and fructose syrup [33].

Rheological method was used by Su and Lannes [34] to predict the fat network formation in ice cream with three types of fats (hydrogenated, low *trans*, and palm fat), before and after the ageing process. The maximum compression force, overrun, and melting profile were determined in the prepared ice creams. The product with hydrogenated fat showed better response of the ageing process than the low *trans* fat product. Greater differences between the three were found. The distinction on structure formation of hydrogenated fat was presented by overrun, compression force and meltdown determinations.

Esteller et al. [35, 36] evaluated the substitution of hydrogenated fat and sucrose in hamburger buns formulas were replaced by polydextrose (Litesse II®), salatrim (Benefat™), and sucralose (Splenda®) and their effects on crust color. The results showed that these ingredients can be used as an option to produce low calories baked products.

In the sensorial analysis, the special chocolate filling confectionery, diet + light, produced with Benefat®, as a fat substitute, obtained high levels of purchase intention and thus can be considered as a great product from a market potential point of view [37].

The potential use of beta-glucans as hydrocolloids in the food industry is based mainly on their rheological characteristics, i.e., their gelling capacity and ability to increase the viscosity of aqueous solutions [38]. Oat extract to replace hydrogenated vegetable fat to prepare cakes was used by Rios and Lannes [39], and the structural quality characteristics and shelf life were evaluated; the characteristics of cakes increased without losses of quality.

Inulin was tested as a fat substituted by Richter and Lannes [37] and Laguna et al. [40] in fillings and biscuit, respectively. Fat replacement with inulin or hydroxypropyl methylcellulose provided acceptable biscuits until determined limit. In fillings a good acceptance was obtained also.

Vegetable fat was partially replaced by potato maltodextrin and specially derived waxy-maize maltodextrin aqueous gels in different proportions, in filling formulations. The increase in the amount of fat reduction resulted in an increase in hardness and change in color of the final product, but acceptable by consumers [41].

5.3. Characterizing fat components in emulsions

Water in oil (W/O) emulsions have continuous phase made up of combinations of fats and oils. Butter and margarines, for example, are solid in the refrigerator (around 5°C) and almost liquid above 40°C. Unfortunately, when butter is removed from the refrigerator, there are some problems. Temperature sweep tests at a fixed frequency and strain rate can allow these products to be characterized as a function of temperature and specific melting point. This can help determine the combination of fats and oils required or fat substitute to obtain optimized spreading and texture.

Products containing more crystallizing fat melt a bit slower than products containing more oil. Softening of fresh cheese products rely on a larger extent on dispersion of the product in saliva, although the product consistency reduces considerably as a result of the melting of fat under mouth temperature. A completely molten fresh cheese will be has much more residual thickness in the mouth than margarine.

As for emulsions with plastic rheology, spreading does not tend to change the emulsion structure dramatically unless the volume fraction of the dispersed phase is very large and spreading induces coalescence of the emulsion droplets. Low-fat spreads may coalesce unless the water droplets are structured with biopolymers to retard re-coalescence under shear.

Processing in the mouth will result in more dramatic microstructural changes. First, consider the behavior of a food emulsion in the mouth during mastication when it is eaten pure, i.e., not on bread. In this situation, a typical food emulsion undergoes a chain of events, although some stages may occur simultaneously or may be unnecessary for a specific food emulsion.

The breakup of a food emulsion can be very important for flavor release. For soluble flavors, in both the lipid and the water phases, diffusion should be fast enough to allow the flavors to move from the product to the saliva within the typical residence time in the mouth.

For studying ice cream formulations using different types of fat, rheology provides information to analyze the effect of each fat studied in the ice cream structure [42, 43].

6. Final considerations

Despite of the calories, maybe of some undesirable triacylglicerols and fatty acids, fats and oils occur naturally in a wide variety of sources for providing distinguish material. Hundreds of seeds and oily fruits, animals, and marine sources supply oils. Many food products can be produced using these oils. However, the production of low-fat food is also of great interest due to the current consumer market. Healthiness and beauty are common interests between people nowadays. Nevertheless, the technology for production of low-fat products cannot be overlooked.

Author details

Suzana Caetano da Silva Lannes* and Alline Aurea do Amaral

*Address all correspondence to: scslan@usp.br

Pharmaceutical-Biochemical Technology Department, Pharmaceutical Sciences School, University of São Paulo, São Paulo, SP, Brazil

References

[1] Agência Nacional De Vigilância Sanitária-Anvisa. RDC n.270, 22 September 2005. Technical Regulation for oils and fats. http://www.anvisa.gov.br/alimentos/legis/especifica/regutec.htm#o (accessed March 2015).

[2] O'Brien, R.D.; Fats and Oils: Formulating and Processing for Applications. 3rd ed. Boca Raton: CRC Press, 2009. 574p.

[3] Sato, K. Crystallization behavior of fats and lipids: a review. Chemical Engineering Science 2001, 56: 2255–2265.

[4] Marangoni, A.G.; Wesdorp, L.H. Viscoelastic properties of fats. In: ______, eds. Structure and Properties of Fat Crystal Networks. 2nd ed. Boca Raton: CRC Press, 2013. cap.5, p. 147–158.

[5] Herrera, M.L.; Falabella, C.; Melgarejo, M.; Anon, M.C. Isothermal crystallization of hydrogenated sunflower oil. 1. Nucleation. Journal of the American Oil Chemists' Society 1998, 75: 1273–1280.

[6] Timms, R.E. Crystallization of fats. In: Hamilton, R.J., ed. Developments in Oils and Fats. New York: Chapman & Hall, 1995. cap.8, p. 204–223.

[7] Rousseau, D. The microstructure of chocolate. In: McClements, D.J., ed. Understanding and Controlling the Microstructure of Complex Foods. Boca Raton: Woodhead Publishing, 2007. cap.24, p. 649–690. (Woodhead Publishing in Food Science, Technology and Nutrition).

[8] Kaneko, F. Polymorphism and phase transitions of fatty acids and acylglycerols. In: Garti, N., and Sato, K. eds. Crystallization Processes in Fats and Lipids Systems. New York: Marcel Dekker, 2001, p. 53–98.

[9] Calorie Control Council. http://www.caloriecontrol.org/articles-and-video/feature-articles/glossary-of-fat-replacers (accessed March 2015).

[10] McClements, J.; Demetriedes, K. An integrated approach to the development of reduced-fat food emulsions. Critical Review in Food Science and Nutrition 1998, 38: 511–536.

[11] Jarvis, N., O'Bryan, C. A., Ricke, S. C., Crandall, P. G. The functionality of plum ingredients in meat products: a review. Meat Science 2015, 102: 41–48.

[12] Rocha, J. C. B.; Lopes, J.D.; Mascarenhas, M. C. N.; Barrera-Arellano, D.; Guerreiro, L. M. R.; Cunha, R. L. Thermal and rheological properties of organogels formed by sugarcane or candelilla wax in soybean oil. Food Research International 2013, 50: 318–323.

[13] Freitas, L.C.; Monte, A.D.M.O.; Cavalcante, T.A. Mercado de hidrocolóides no Brasil. Revista de Química Industrial 1996, 64: 708–709.

[14] Katzbauer, K. Properties and applications of xanthan gum. Polymer Degradation and Stability 1998, 59: 81–84.

[15] Glicksman, M. Hydrocolloids and the search for the "Oily Grail". Food Technology 1991, 45:.94–103.

[16] Laneuville, S.I.; Paquin, P.; Turgeon, S.L. Formula optimization of a low-fat food system containing whey protein isolate-xanthan gum complexes as fat replacers. Journal of Food Science 2005, 70: S513-S519.

[17] Ward, F.M. Hydrocolloid systems as fat mimetics in bakery products: icings, glazes and fillings. Cereal Foods World 1997, 42: 386–390.

[18] U.S. Food and Drug Administration. www.fda.gov (accessed March 2015).

[19] Cunha, P.L.; Paula, R.C.M.; Feitosa, J.P.A. Polissacarídeos da biodiversidade brasileira: uma oportunidade de transformar conhecimento em valor econômico. Química Nova 2009, 32: 649–660.

[20] Almeida, P. F.; Lannes, S. C. S. Application of gelatin obtained by product of chicken in chocolate spread. In: XIX Semana Farmacêutica de Ciência e Tecnologia, 2014, São Paulo. Brazilian Journal of Pharmaceutical Sciences 2014, 1: 1–1.

[21] Amaral, A.A.; Lannes, S. C. S.; Feitosa, J. Influence of the addition of gums on the rheological behavior of fillings for chocolates. In: 2º Congresso Brasileiro de Reologia, 2013, Aracaju. Livro de Resumos. Aracaju: Associação Brasileira de Reologia 2013, 1: 56–56.

[22] Pinto, M. M. M.; Mello, K. G. P. C.; Polakiewicz, B.; Lannes, S. C. S. Uso de quitosana modificada como substituto do óleo em formulações de emulsão tipo maionese. In: XII Semana Farmacêutica de Ciência e Tecnologia da FCF-USP, 2007, São Paulo. Brazilian Journal of Pharmaceutical Sciences 2007, 43: 18–18.

[23] Rahmati, N.F., Tehrani, M. M., Daneshvar, K., Koocheki, A. Influence of selected gums and pregelatinized corn starch on reduced fat mayonnaise: modeling of properties by central composite design. Food Biophysics 2015, 10: 39–50.

[24] Deman, J.M.; Deman, L. Specialty fats based on palm oil and palm kernel oil. Malaysian: Malaysian Palm Oil Promotion Council, 1994. (Palm oil information series).

[25] Lipp, M.; Anklam, E. Review of cocoa butter and alternatives fats for use in chocolate. Part A. Compositional data. Food Chemistry 1998, 62: 73–97q1.

[26] Craig-Schmidt, M.C.; Teodorescu, C.A. *Trans* fatty acids in foods. In: Chow, C.K., ed. Fatty acids in foods and their health implication. 3rd ed. Boca Raton: CRC Press; London: Taylor & Francis, 2008. cap.15, p. 377–437. (Food Science and Technology; Food Science and Technology).

[27] Martin Jr., R.A. Chocolate. Advances in Food Research 1987, 31.

[28] Petrauskaite, V.; De Greytw, K. M.; Huyghebaert, A. Physical and chemical properties of *trans*-free fats produced by chemical interesterification of vegetable oil blends. Journal of the American Oil Chemists' Society 1998, 75: 489–493.

[29] Karabulut, I.; Turan, S.; Ergin, G. Effects of chemical interesterification on solid fat content and slip melting point of fat/oil blends. European Food Research and Technology 2004, 218: 224–229.

[30] Ribeiro, A.P.B.; Moura, J.M.L.N.; Grimaldi, R.; Gonçalves, L.A.G. Interesterificação química: alternativa para obtenção de gorduras zero *trans*. Química Nova 2007, 30 : 1295–1300.

[31] American Cancer Society. http://www.cancer.org/healthy/eathealthygetactive/takecontrolofyourweight/low-fat-foods (accessed 30 March 2015).

[32] Nebesny, E.; Zyzelewicz, D. Effect of lecithin concentration on properties of sucrase-free chocolate masses sweetened with isomalt. European Food Research Technology 2005, 220: 131–135.

[33] Silva Junior, E.; Silva, E.R.T.; Muramatsu, M.; Lannes, S. C. S. Transient process in ice creams evaluated by laser speckles. Food Research International 2010, 43: 1470–1475.

[34] Su, F.; Lannes, S. C. S. Rheological evaluation of the structure of ice cream mixes varying fat base. Applied Rheology 2012, 22: 63871–63877.

[35] Esteller, M. S.; Lima, A. O.; Lannes, S. C. S. Color measurement in hamburger buns with fat and sugar replacers. Lebensmittel-Wissenschaft+Technologie/Food Science +Technology 2006, 39: 184–187.

[36] Esteller, M. S.; Amaral, R. L.; Lannes, S. C. S. Effect of sugar and fat replacers on the texture of baked goods. Journal of Texture Studies 2004, 35: 383–393.

[37] Richter, M.; Lannes, S. C. S. Bombom para dietas especiais: avaliação química e sensorial. Ciência e Tecnologia de Alimentos 2007, 27: 193–200.

[38] Rios, R. V.; Pessanha, M. D. F.; Almeida, P. F.; Viana, C. L.; Lannes, S. C. S. Application of fats in some food products. Food Science and Technology 2014, 34: 3–15.

[39] Rios, R. V.; Lannes, S. C. S. Evaluating of quality characteristics of cakes with different fat replacers. In: XIX Semana Farmacêutica de Ciência e Tecnologia, 2014, São Paulo. Brazilian Journal of Pharmaceutical Sciences 2014, 1: 1–1.

[40] Laguna, L.; Primo-Martín, C.; Varela, P.; Salvador, A.; Sanz, T. HPMC and inulin as fat replacers in biscuits: sensory and instrumental evaluation. LWT—Food Science and Technology 2014, 56: 494–501.

[41] Hadnađev, M.; Hadnađev, T. D.; Dokić, L.; Pajin, B.; Torbica, A.; Šarić, L.; Ikonić, P. Physical and sensory aspects of maltodextrin gel addition used as fat replacers in confectionery filling systems. LWT—Food Science and Technology 2014, 1: 495–503.

[42] Lannes, S. C. S.; Silva Junior, E.. Viscoelastic properties of mixtures for ice cream. In: Geoffrey Mitchell (Org.). Rheology: Theory, Properties and Practical Applications. 1st ed. Novapublishers, 2013, 1: 01–27.

[43] Lannes, S. C. S.; Ignacio, R. M.. Structuring fat foods. In: Innocenzo Muzzalupo. (Org.). Food Industry. 1st ed. Rijeka—Croatia: Intech Open Science, 2013, 1: 65–91.